THE MAN WHO CAUGHT THE STORM

THE LIFE OF LEGENDARY
TORNADO CHASER TIM SAMARAS

BRANTLEY HARGROVE

Simon & Schuster
New York   London   Toronto   Sydney   New Delhi

Simon & Schuster
1230 Avenue of the Americas
New York, NY 10020

First Simon & Schuster hardcover edition April 2018

SIMON & SCHUSTER and colophon are registered trademarks of Simon & Schuster, Inc.

For information about special discounts for bulk purchases, please contact Simon & Schuster Special Sales at 1-866-506-1949 or business@simonandschuster.com.

The Simon & Schuster Speakers Bureau can bring authors to your live event. For more information or to book an event, contact the Simon & Schuster Speakers Bureau at 1-866-248-3049 or visit our website at www.simonspeakers.com.

Interior design by Carly Loman

Manufactured in the United States of America

10 9 8 7 6 5 4 3 2 1

Library of Congress Cataloging-in-Publication Data is available.

ISBN 978-1-4767-9609-3
ISBN 978-1-4767-9611-6 (ebook)

For Kathy and Laurie

CONTENTS

Prologue   *1*

PART ONE

PART TWO

PART THREE

PROLOGUE

THE FIRE DEPARTMENT'S siren sounded over Jarrell, Texas, just after 3:30 p.m.—a shrill, oscillating note, like an air-raid alarm, swelling and fading and swelling again. It filled every street and pressed in through the windows of every house in the little Czech farming town some forty-five minutes north of Austin. The siren was only ever used to call volunteers to the station, but anyone who had been paying attention that afternoon knew this time was different. They had ten minutes, maybe twelve at the outside.

The TV meteorologists were already tracking the tornado near Prairie Dell, four miles to the north of Jarrell. It writhed like a coach-whip snake at first, its thin form roping liquidly over prime farmland, where the Hill Country gives way to Blackland Prairie. For a time it seemed to track neither north nor south, but to trace a languid orbit in place.

Then came the shift.

The graceful ribbon was suddenly gone, replaced by this other thing—an inchoate, gray miasma, not so much a tornado as a wall of smoke, leaching up out of the earth. The twister was on the move now, bearing southwest on a collision course with Jarrell.

When it finally appeared on the northern horizon at about 3:40

1

p.m., it was a sight townspeople would resummon in their dreams for years to come. Bristling with debris and blackened with rich soil scoured from the fields, it looked ancient and immutable, its sooty wings spread wide. What it looked like was the end of the world.

The tornado was as wide as thirteen football fields laid end to end. Little Jarrell could have disappeared inside it, swallowed and gone. But the town that day was largely spared. The darkness passed to the west, away from the most densely populated parts of town. The trouble was, there were still people in its path, and more than usual on a Tuesday afternoon. School had let out for summer on the Friday before.

Double Creek Estates, a collection of modest single-story, wood-frame-and-brick starter homes, was huddled in the lowlands just north-west of downtown. Like most everywhere else in Texas, the houses didn't have basements; the water table was too high, and the limestone bedrock too shallow. With no choice but to shelter aboveground, the residents did as they had always been instructed: they sought out hall-ways, bathtubs, closets. They gathered their children into these spaces and listened to the wind, then the breaking glass, the groaning wood, and finally the raw sound of it, a deafening, toneless static, until the roofs and the walls surrounding them fell away.

To say that the neighborhood was flattened would be to imply that there were ruins left to pick through in search of survivors. For National Weather Service surveyors cataloging the indexes of destruction, the tornado was notable for how few injuries it produced: one serious and ten minor along the periphery of the path. All else was fatality. Aboveground, inside the core, the odds of survival were nearly zero. The Hernandez family was the outlier—they lived only because Gabriel and his wife had insisted on carving a belowground shelter out of the limestone by hand.

Their home and some thirty others had not simply been razed; the foundations had been scraped clean, in some cases even of plumbing. The lawns surrounding them weren't littered with debris. The remainder of the structures—the frame, brick, drywall—had been "granulated," to quote one surveyor, and strewn over long distances downwind.

Eventually the clouds burned away, and the sun shone again. The carcasses of hundreds of cattle, some of them with the hide stripped clean of every follicle, lay mud-plastered in barren fields that were recently green. They were strewn through the woods, where even some of the oldest oaks had been pulled up, root and trunk. Many of the trees that still stood were swaddled in stiff sheet metal blown from nearby buildings and had been denuded and nubbed off at the top, like totem poles. The countryside reeked with decay, as if a suppurating wound had been opened on the land itself. And in a sense, one had.

The Jarrell twister produced what is to this day regarded as the most extreme damage researchers have ever encountered. More than five hundred feet of asphalt had been peeled away from the county roads where the tornado crossed, exposing the crushed-stone bedding beneath. Eighteen inches of topsoil had been suctioned from the lush fields of wheat and cotton, transforming them into vast muddy scars pooled with stagnant water. The people that saw these phenomena understandably wondered how the wind could do such things.

A couple of weeks later, once the debris had been cleared away, and the remains had been recovered as thoroughly as could be expected, my family and I drove slowly among the naked foundations. We lived not far from Jarrell, and during a recent summer, when I was fourteen or so, I had stocked shelves in the town's general store. As the blocks of concrete and tracts of bare soil slipped past, I could have believed that no one had ever lived in Double Creek. For all I knew, this could have been a brand-new development, with fresh slabs still waiting for the frames and anchor bolts.

But the truth was, they had contained lives, and their erasure was the brutal consequence of prolonged exposure to nightmarish winds. I learned later that the tornado had crawled through Double Creek at around ten miles per hour, and sometimes much less. Given its size, that meant some residents were subjected to sustained winds of more than two hundred miles per hour for minutes on end. The tornado became a grinder awash in shards of wood and metal and God knows what else. If the wind didn't kill you, the things it carried would.

In that kind of hell, what can stand? Driving through, where I noted a stray tennis shoe still dangling from a barbwire fence, I couldn't help but wonder, *Why Double Creek?* In the hundreds of square miles of quiet country between Waco and Austin, in all that open farm- and ranch-land where the storm could have spent its fury in harmless isolation, it found this tiny neighborhood—probably less than a mile square in size—and swept it from the world. Around here, the loss felt immeasurable. Entire families were simply gone, including every member of the Igo clan. The fifteen-year-old twin Igo boys, John and Paul, had worked at the same general store as me, where we'd overlapped just once.

Yet outside my little corner of Texas, the missing and the dead were one part of a much larger story. In cold statistical terms, the twenty-seven killed were but a portion of the yearly national toll. On average, tornadoes will claim eighty lives annually, a figure convulsed by significant variance; in 2011, for example, 316 died in a single day. The loss that year was staggering; it wasn't just the people who died but the economic damage the storms incurred: some $28 billion in cost between April and May. In the United States, the damage caused by tornadoes has outstripped that of fires, earthquakes, and floods and is nearly on par with that from hurricanes. The most violent type—the EF4s and EF5s—are exceedingly rare, accounting for roughly one to two percent of all tornadoes. A storm chaser may never see an EF5 in his lifetime. Yet despite their infrequency, some seventy percent of tornado fatalities are attributable to the deadliest breed. The scale of these disasters is nearly beyond accounting.

An EF5 flattened a swath of Joplin, Missouri, on May 22, 2011. The trail of destruction it left behind was some six miles long and up to eighteen hundred yards wide. With 158 dead, and thousands of homes and buildings severely damaged or destroyed, the town called to mind images from Hiroshima and Nagasaki after the bombs. Streets were unrecognizable to residents who had lived on them for years. Joplin, as they had known it, no longer existed.

But the true testament to the tornado's power can't be observed from a helicopter circling high above. It's on the micro level, in the little scenes and dispositions of objects that seem to defy explanation and the laws of physics. Researchers refer to these as "incredible phenomena." In Joplin, damage surveyors found a truck that had been wrapped around a tree, the front bumper in contact with the rear. Cardboard had imbedded into the siding of the high school. A shard of wood had impaled a concrete parking curb. Steel manhole covers had simply disappeared.

The month before Joplin, the small town of Smithville, Mississippi, was largely destroyed by an EF5. A Ford Explorer was lofted for half a mile before smashing into the top of the town's 130-foot-tall water tower. The vehicle then tumbled another quarter mile before finally coming to rest. The same day, in Tuscaloosa, another tornado sheared steel trusses from the train trestle over Hurricane Creek Canyon. One of the trusses, weighing about thirty-four tons, was blown one hundred feet *uphill*. Farther along the tornado's path, at a coal yard, a thirty-six-ton railcar was lifted from its tracks and hurled nearly four hundred feet. Eyewitnesses say it didn't tumble, it flew. On May 20, 2013, an EF5 chewed through the outlying suburbs of Oklahoma City, killing twenty-four. In one subdivision, two brick houses, sitting some ten feet apart, had been pierced by the same two-by-six board. It passed clean through the first before lodging in an interior wall in the second.

Yet for all these brute displays of unfathomable strength, there are just as many stories surrounding the wind's incongruous tenderness. The same day SUVs and railcars were airborne, a restaurant in Ringgold, Georgia, called Chow Time, was damaged beyond repair. The ceiling was caved in, the walls collapsed, but at the center of it all sat a cake fit to be served. Its frail glass container hadn't even cracked; every last swirl of frosting was pristine. In Joplin, a child's play set, made of nothing sturdier than hard plastic, survived unscathed. Surrounded on all sides by uprooted trees and pulverized homes, it was as though it alone had been spared.

With seasonal regularity, the continental United States is scored with the tracks of tornadoes, from the Deep South up to the High Plains. Most last only seconds, a fleeting discontinuity swiftly rectified. But a vanishing few sink lasting roots into the storms that spawn them, becoming so enormous, and so powerful, that they sustain themselves for hours over dozens of ruinous miles. It's one of the most awesome expressions of force in the natural world, and also one of its most unpredictable.

Tornado warnings issued by the National Weather Service have a false-alarm rate of roughly seventy percent. When they are accurate, the average lead time between warning and impact is around fourteen minutes. The people of Joplin had seventeen minutes. On September 20, 2000, Xenia, Ohio, received no warning at all. The weather service had issued a severe-thunderstorm warning, but its radar failed to detect the signature of the coming tornado. By the time the F4 arrived, it had knocked out power to four of Xenia's five sirens. One man was killed when a tree crushed his car, and more than one hundred others were injured. In 2016, the Storm Prediction Center, the government's finest severe-weather forecasting unit, issued a morning outlook that indicated a two percent probability of tornadoes across northern Missouri, at around lunchtime. By that afternoon, it became clear its forecast had missed the signal in the noise: one of the largest August tornado outbreaks ever recorded was under way in Indiana and Ohio.

The tornado's appearance each year is inextricable from human memory in North America and across the earth's temperate climates. The Arikara called it the Black Wind. To others it was Whirlwind Woman, the bringer of life-giving rain and death, her arrival as inevitable as the seasons. The Cheyenne and Arapaho believed the caterpillar created the whirlwind, and entomologists know today that they often hatch when the barometric pressure falls before a storm. The Plains tribes grasped something essential about the correlation, though the signs to them were supernatural.

Those who've been caught by surprise couldn't be faulted for concluding that, even with all our Doppler radar arrays and rawinsonde weather balloons, meteorology hasn't demystified tornadoes one bit. We now live in an era when the *Mars Pathfinder* rover has touched down on the Red Planet. The human genome has been mapped. Physicists use particle accelerators to study the subatomic fundament of all matter. But for all our technological achievement, twisters still have the power to confound even the most advanced civilization the planet has ever known. When they come, the best any of us can do is get out of the way or place layers of thick steel and reinforced concrete between our bodies and the wind.

We still can't predict which storms will give rise to killer twisters, nor do we fully understand the process by which weak tornadoes become strong. Until recently, the core of a tornado remained as remote and untouchable as the surface of the sun. Researchers still dream of the day when forecasters have the tools to issue thirty-minute advance notifications—warnings not only of imminent tornado formation, but of potential intensity. Such notice could give residents in towns such as Jarrell the time to shelter in place, to find adequate protection, or to escape. But before that day comes, scientists must first answer questions about the nature of the vortex so fundamental they could be posed by a layman: Why do some storms produce tornadoes and others do not? What sequence of events gives birth to the biggest, long-track EF5s? Can we identify the signs before it's too late?

There is a future in which there are answers to these questions and we are given more than fourteen minutes between the warning and the darkness at the door.

Humans have always invented monsters to explain the inexplicable. In Romania, where tornadoes are infrequent but have been known to occur, folk mythology tells of the *balaur*, or dragon. As the myth goes, this creature soars high above, trailing a long, lashing tail. Its roar is deafening. Its breath turns the water in the clouds to ice. It carries people into the sky. And in the end, the dragon leaves nothing but ruin in its wake.

Hundreds of years ago, the Romanian villagers couldn't explain what had happened to them, so they invented a story. The same happened nearly everywhere else tornadoes touched down. The world has changed enormously since these stories were first uttered, but not every myth has faded.

The tornado is one of the only real dragons the modern world has left. And the only way to dispel the frightening unknown is for someone to steal away with its secrets. If we are to glimpse what lies at the heart of one of the planet's greatest mysteries, someone must first journey to a place few have witnessed and live to tell about. The task requires invention, a tolerance for danger, and an unusual breed of talented tinkerer.

We know this because there once was such a man.

He got closer than anyone before or since, committing his life with fanatical devotion to the chase and a search for answers. He wasn't the decorated expert you'd expect. He wasn't an eminent scientist with a fine academic pedigree or the resources of a major research institute. For the most part, he was just a regular guy—with a dream, an uncommon set of skills, and an insatiable appetite for tracking down extreme storms. This man set out on a mission he'd been told was highly inadvisable, if not completely suicidal. To the shock of the scientific community, he pulled it off. He swept the shroud aside, if only briefly, and showed us all something we'd never before seen—the heart of the tornado.

By proving that the core was not untouchable after all, he pushed the field forward. Then, he pushed too far. He spent decades searching for the ultimate storm. And when he found the one, it upended everything he thought he knew. This is the story of a man who caught what he was chasing. He surpassed his peers, seeming to transcend science with his ability to read the vortex—until the object of his fascination finally turned. It swept down from the sky, carried him up, and took his life.

PART ONE

CHAPTER ONE

THE WATCHER

July 21, 1993

F OG CLINGS TO the low swells of eastern-Colorado rangeland as dawn breaks. The mist walls off the far horizon, and for a few short hours the high plains feel a little more finite. The still air is cool and heavy, almost thick enough to drink. This is how these days often begin. The atmosphere is primed, the air a volatile gas. All it needs is a match.

By noon the summer sun presses down. The sagging fog begins to heat, and the dense haze disperses, along with the morning chill. The faultless dome of heaven takes on a hard, lacquered blue, and the windless air stirs as a steady breeze sweeps up out of the southeast. A skein of fluffy clouds sets adrift on the horizon.

The average citizen sees a sunny day. Tim Samaras sees a fine afternoon for a tornado.

In all likelihood, Tim has been tracking this setup and planning his chase for days. He's already en route to the plains from his home in suburban Denver. As the sun reaches its peak, his hail-battered Datsun pickup enters the storm chaser's cathedral. There is none of the verticality of trees or mountains here to modulate the wind or break his sight lines. Once the sheltering Front Range fades from the rearview mirror, he's naked to the lungs of the earth, in an unadorned country where the passage of miles can feel more like a few hundred yards.

Today looks fit for a picnic. The wind is picking up, but temperatures are mild, edging into the upper seventies. If he were to stand outside long enough, he'd probably get a sunburn. But Tim sees fair weather through uncommon eyes. He sees rain and wind and potential violence in an untroubled sky. While others bathe in the rays, Tim waits and watches the scattering of clouds. The atmosphere, he knows, is drinking in the sun's radiation, like a drunk about to get mean.

By midafternoon, the benign, almost friendly looking clouds begin to curdle, and the fields beneath them darken. When the feathered tops of fluffy cumulus suddenly take on a polished hardness and start building toward the troposphere like columns of wildfire smoke, his chase begins. The day's first storm pops less than an hour east of his home in Lakewood, Colorado, but it doesn't distract him. He's more interested in the storm that doesn't exist yet. Chasing is prognostication and timing. It's predicting where the tornado could happen, and being there at the precise moment that it does. His guide is a subtle map of invisible boundaries—diffluent and converging rip currents of air, surges of southeastern moisture, western aridity, and polar cold. The clues point farther east, to the place where the morning fog has lifted and the wind now freights vapor over a parched, cloud-shaded prairie. The blue Datsun picks its way along a lonely stretch of US Route 36. Tim scans the storm towers, listens to the squawk of the weather radio, and waits for the magic hour.

As the sun settles low behind him, a storm to the southeast catches his eye. It's isolated, rising above the cloud deck like a mountain peak surrounded on all sides by foothills. The storm appears to be closing in on a tongue of moist, unstable air, which might as well be a stand of drought-killed trees before a forest fire. Tim leaves the highway and steers south. By 7:45 p.m. he parks along a gravel road in the remote ranch country, somewhere south of a little town called Last Chance. The rain is coming down hard now, so he keeps to the cramped cab of the Datsun into which he has tucked his lean, five-foot-seven-inch frame. He feels the truck beneath him rocking gently. He watches dingy curtains of soil and water strafe across tufts of buffalo grass and grama. The storm has come to him.

At a minute before 8:00 p.m., there isn't another soul in sight except for the monster that is just now emerging from the dust and darkness. Tim has the big show all to himself, and what a show. This might be shaping up to be the biggest storm of his chasing career thus far. An anvil cloud never fails to evoke in his mind the pyrocumulus of an atomic bomb, its head a tumescent riot of burls and bulges. He could spend hours letting his eyes roam across every pulsing shred of cloud, every rotating square foot throwing shadow over miles of high plains. The storm's edifice luminesces every other moment like a paper lantern. By the end of its life cycle, the tower of electricity and ice and hurricane-force gale will have dispersed the energy equivalent of a twenty-kiloton nuclear warhead. This storm isn't anywhere near finished yet. Tim is no meteorologist, but he knows a long-tracker when he sees one.

Though the uninitiated could easily mistake the darkness beneath the cloud for rain, he would be gravely mistaken. All around, the storm is alive, pulsing, almost animate, yet the shadow beneath appears depthless, without feature or flux. If there's one thing Tim has learned, it is that looks lie in Tornado Alley. Behind the dark wall, there's more than just rain. Hiding inside, Tim's now certain, is a serious tornado.

With one hand, Tim focuses the camcorder on the eastern horizon, and with the other, he lifts the handset to his mouth.

"WJ0G," he says, his voice raised slightly over the sharp drum of rain against the roof. A burst of static follows from the ham radio.

"G, go ahead."

"Okay, I'm going to have to confirm it's still on the ground. The whole base is rotating, and the base is almost touching the ground . . . the whole rotating wall cloud."

"WJ0G, N0LVH confirm. . . . Can you provide a location?"

"Okay, it's definitely south of Lindon, and it's almost directly to my east now. I'm probably five miles south of the highway."

Tim isn't a newbie—nor is he an old hand yet—but his forecast today is as spot-on as a professional's. And like a pro, he knows well enough to give this one a wide berth. Getting too close to a rain-wrapped tornado is a bit like piloting a dinghy into a blind fog. Usually, the craft sails through without incident. But there's always a chance that the dinghy takes the bow of an oil tanker broadside. The tornado behind the veil is the one you never see coming.

Fortunately, and by some stroke of impossible luck, no chaser has ever died this way. For the watchers like Tim, there is simply no good reason to get that close. Besides, he prefers to look at the storm in all its mammoth totality, its increasing scales of movement like the turning of gears. He also understands that the tornado isn't the thing, it's the consequence, the eddy in the stream. It exists on both visible and invisible spectra. Its behavior is governed by forces too complex and too grand to track in real time. The day a chaser forgets this may be his last. There's only one rule out here: Never get too close or too cocky. Never be too sure you've seen the worst the storm can deliver. The sky can always show you something you haven't seen yet.

With a few years of chasing experience, Tim has taken the rule to heart.

He is a compact man, lean even as a father of three at the age of thirty-five. He has thickly tendoned forearms shot through with heavy veins, and the rough hands of a mechanic. His most defining facial characteristic is a prominent nose, which at first proceeds gently from his face before plunging at the bridge, lending him a hawkish, though not-at-all unfriendly appearance. His dark hair is beginning to recede ever so slightly at the temples, but its retreat makes him look more distinguished. These days he wears a beard, thick and glossy.

A Coloradan with a Greek Albanian father and a Hispanic mother, Tim has a westerner's reflexive congeniality; he will pause, even at the most inopportune moment of the chase, to provide the locals who approach him with pertinent meteorological updates.

He has never chased outside his home state and plans to make a pilgrimage to Texas soon. The Lone Star State is hallowed ground where the real fanatics go, as a friend puts it, "to see the hand of God passing over the earth." It will be the first, Tim hopes, of many such trips to come. An initiate into this wandering brotherhood, he has taught himself how to read a weather map and how to identify the morphological features of storms. Tim believes his apprenticeship under the tamer Colorado skies has prepared him for the kind of tempest that belongs to history, the kind spoken of with reverence. Usually, they're known by the tacit shorthand of location and, if there's been more than one, a date. Mention Lubbock, Texas, to a chaser—or the great Tri-State twister; or Andover, Kansas; or Woodward, Oklahoma—and nothing else need be said by way of explanation.

Witness one of the giants, and the outside world—the life at home and all its worries—falls away. Maybe it never fully comes back. Nothing else will ever feel quite like standing inside the lungs of a storm. As long as a chaser lives, he won't forget the way the inflow presses at his back; or the grit sandblasting his calves; or the air itself, which carries a latent charge; or thunder like the report of an 800 mm railgun caroming off the clouds.

Every time he chases, even if he misses the tornado, Tim learns. He's good, and someday soon he might become great. He's honing his forecasts, which are as much an art as a science. He's mastering the way the most revered chasers channel their awe, fear, and adrenaline into a useful kind of hyperawareness.

This moment near Last Chance is cause for celebration. For this, Tim would gladly pay a dozen busts and near misses. For this, he'd drive any number of miles.

The tornado shows itself now, emerging briefly from behind its curtain of rain. It isn't one vortex, but *several*, each briefly resolving against the horizon, where the edge of the storm meets the day. It isn't the iconic funnel so much as a shaft through which vortices move, each narrow and often transitory, yet deceptively violent.

Just as quickly as it emerged, the vortex disappears again. The storm drags a broad slug of rain across the intervening fields, and there is little else to see now apart from atomized water, blowing dust, and the storm anvil splayed for miles above Tim's head. For now, he remains in the Datsun, keeping his distance. He could get closer and angle for a clearer view—he could try to pierce the rain veil and punch the hail core—but he doesn't. Not today, not with this tornado.

There are scientists out there who once ventured behind the veil, who didn't watch so much as hunt. They sought knowledge, not thrills. They roamed the plains with purpose, or at least they used to. He saw them in a documentary a few years back, and their exploits are the reason he's out here now. It's why he chases, though he probably couldn't even articulate for himself all of the reasons their work has touched him so. He isn't an atmospheric scientist like them; he didn't even go to college. He's just a watcher waiting for the rain to pass.

Within another five minutes, he steps out onto the gravel. The wind whips through his legs. Behind him, the sun breaks through the western clouds, and in the light, the storm blanches from carbon gray to an immaculate white.

"My God," he says.

The storm is vast. It's like looking up at the mountains west of his home, only the anvil soars far higher. The bulbous cloud has been smoothed and hardened by the wind like the jagged face of the Front Range. With the sun at his back, the contrast is weak, but he can just make out the pale suggestion of a single vortex a mile or so out. The tornado is both mirror and crystal: it reflects or refracts depending on the viewer's position relative to the sun. Dark as granite when backlit, it can be wan as milk with the light behind the viewer. The funnel can mimic the verdure of the crops beneath, or the brown-black of soil.

Tim isn't sure whether he believes in a higher power. Whether the sight before him is the work of God or a simple disequilibrium, it is undeniably a magnificent work to behold. Yet as much as the tornado invites a sense of mysticism, Tim tries to resist the impulse. Back in the real world he's an engineer of an odd sort; he places his faith in the tactile, the quantifiable. What separates him from most chasers is that he believes everything on this planet—including the godlike mystery on the plains before him—can be measured. He wagers that, with the right tools and steady nerves, even the deepest unknown can be plumbed and understood.

Today is not the day to test this belief. The storm is too dangerous, and Tim has much more still to learn. But his curiosity is boundless, and at Last Chance there is some part of him that is plotting, imagining how to part the rain, to pierce the core, to touch the tornado's flux and fury.

CHAPTER TWO

A BOY WITH AN
ENGINEER'S MIND

A s a kid, it wasn't enough for Tim Samaras to see that the gadgets
around him worked. He had an irrepressible need to know *how*
and *why*. The bane of his mother's household appliances, he dismantled her blender to see why the blades spun so fast. At ten years old, he
autopsied the television set in an attempt to determine how colors and
shapes flashed across the screen. That these things worked perfectly
fine before he took them apart was not something he seemed capable
of taking for granted.

Rather than reprimand the boy, his parents gave in to his tinkering. His father, Paul—to save himself repeated trips to Sears—kept Tim
supplied with cast-off junk to deconstruct. He went so far as to take out
an ad in the *Rocky Mountain News*, seeking used electronics. So long as
the gadget was free, Paul Samaras would show up at your doorstep, a
salvager combing the Denver suburbs for the benefit of a little boy with
an engineer's mind. Mostly he returned to Tim with antique radios—
the kind with the big dials—clutched under one arm.

Tim's bedroom was his laboratory, and a hazard to bare feet, strewn
as it was with transistors and diodes and capacitors. Here, he brought
the silent radios back to crackling life. Though Paul was a stern and

authoritarian presence in the house, he actively abetted Tim's hobbies. The father had always wanted to become a ham radio operator. The trouble was, he never got around to passing the code tests required by the FCC. Tim found his father's manuals, studied them, and became a licensed amateur at twelve years old, call sign WN0JTV. He built his first transmitter using the horizontal output tube from an old television set. The accomplishment filled Paul with pride, and he soon erected a used two-and-a-half-story Hy-Gain antenna tower next to the house in Lakewood, to amplify the signal his son could receive and broadcast. Since the yard was small, with little room to bury radial wires to ground his antenna, Tim would occasionally sneak out of his bedroom window at night and excavate small trenches in the neighbors' yards for burying his lines. The two-thousand-volt transmitter emitted such a powerful signal that it often infiltrated the electronic organ next door. The little neighbor girl would be in the middle of practice when an awful shrieking and hissing would pour through the organ's speakers, provoking her mother to fits of obscenity.

On special nights, though, Tim's radio fell silent, and the family gathered for one of its favorite films. Paul would drag the dinner table into the living room, and Tim's mother, Margaret, would serve supper in front of the TV. On one such Sunday at six o'clock, six-year-old Tim took his seat next to his brothers, Jim and Jack. The house echoed with the roar of the Metro-Goldwyn-Mayer lion, then the first strident chords of *The Wizard of Oz*.

Tim was entranced by the film—but not by Day-Glo Technicolor, or the timeless parable. Sepia-toned Kansas was what rooted him in his chair that evening. He couldn't take his eyes off the tornado as it roped over the fields toward Dorothy and Toto. A skirt of sod swirled around the twister. The wind howled, the windows shattered, and Dorothy's little farmhouse took flight. Tim could scarcely believe such things existed on the plains east of his home. He would watch the opening moments again and again throughout his childhood. The rest of the film made him sleepy—even the Munchkins and those terrifying flying monkeys. The fantasy didn't hold the same magic he beheld in that vision of raw power and menace.

Three years later, Tim caught his first real glimpse of Dorothy's tempest, from his own backyard. It wasn't the awe-inspiring image he had hoped for, just a small funnel cloud, never in contact with the ground. Still, he sprinted into his neighbor's yard and mounted the swing set, angling for a better view. The thin, introverted boy clutched the metal bars and craned his neck as this snake in the sky undulated languorously over the city in the storm's half-light. It was one of the most beautiful things he'd ever seen.

As he grew, whenever his dual interests in technology and severe weather could align, Tim's eyes would alight. When storms blew over the Rockies, he'd run wire from his window to a power pole outside, attempting to conduct the ambient electrical charge to a lightbulb. When the storm was far off, he'd tune his radio to the dead space between stations and press his ear against the speaker. If he was quiet, he could hear static crashes, the whispers of distant lightning in pulsed white noise.

By the time he was old enough to drive, he'd take his 1967 Ford Galaxie 500, "a huge frickin' boat," his brother Jim says, and park it at the make-out spot on an outcropping near Red Rocks. Tim wasn't a bad-looking young man: slight of build, with a thick head of dark, nearly black hair, and a fine olive complexion. But he didn't come here for the girls. He had simply outgrown watching storms in his backyard. From his perch at Red Rocks, the clouds were practically right on top of him, racing close above the giant amphitheater of pink sandstone.

For spending cash, he clocked after-school hours at an electronics repair shop, hunched over busted CBs, two-ways, and FM radios. The idle tinkering in his bedroom had evolved into a knack for figuring out what was wrong with all range of gadgetry. Before he graduated from Alameda High School in 1976, he was managing the shop. With his skills there was no telling what doors might open were he to attend college. His parents even offered to pay. Their only condition was that he live at home and make his marks. But Tim had no interest in school; he couldn't tolerate sitting still at a desk. He believed he could teach himself anything he needed to know.

It was this imperturbable faith in his own faculties that led Tim to Larry Brown's office in 1978. Brown headed up an outfit of explosives experts at the Denver Research Institute, an applied-engineering firm housed in a trio of boxy, formalist buildings atop a former army barracks on the University of Denver campus. Brown had placed an ad seeking an instrumentation engineer. The gig entailed working with state-of-the-art electronics designed to quantify the destructive force of military ordnance—bombs, that is—among other highly explosive odd jobs. Tim approached Brown's desk wearing jeans that were ripped at the knees, and a T-shirt. He projected an aura of easy confidence, but he wasn't cocksure.

"I'd like a job," he said.

"Well," Brown replied, "I need a résumé."

Tim had brought along no such thing, nor had he ever drawn one up. He told Brown he'd be back, and the next day, sure enough, he walked in with a yellow sheet of paper summarizing, in handwriting, his limited experience. Brown had to admire Tim's gumption. The kid had brass, but no experience beyond working in a mom-and-pop radio-repair shop. Yet here he was, twenty years old, with no college education, and he's knocking on the door of one of the premier research contractors in the West. Despite the fact that Tim was staggeringly unqualified on paper, Brown saw something in the young man. He couldn't explain it, but he could look past the thin résumé to what Tim might become. Brown had been building research organizations for years; he prided himself on spotting talent and had long ago learned to listen to his gut. He offered Tim the job.

The young technician was given little time to acclimate at DRI, and as it turned out Tim didn't need it. A quick study, he taught himself as much electrical engineering and physics as was demanded by his duties: namely, testing and exploding weapons systems. Before he was old enough to drink, he had already earned himself a Pentagon security clearance.

Among his earliest projects was a mammoth conventional-explosives test, which was to be the largest of its kind ever attempted. At White Sands Missile Range in New Mexico, not far from the Trinity Site, where the world's first atomic bomb was detonated, DRI had some 4,440 tons of ammonium nitrate explosive and fuel oil (ANFO) rigged to blow. Under the aegis of what was then known as the Defense Nuclear Agency, the Cold War–era test was intended to simulate the blast of a nuclear weapon.

To measure the blast wave, which would require detail right on down to the millisecond, the agency had contracted with Brown's posse of hotshots to handle the high-speed instrumentation. Brown's techs were consummate MacGyvers, accustomed to dreaming up weird solutions to complex problems. To track the blast's cratering characteristics and the flight of ejected rubble, they buried bowling balls stuffed with flares. At T-minus one second, the flares would ignite, allowing cameras to trace their smoking arcs across the sky.

Tim's role was to manage the more than one hundred ultra-high-speed cameras arrayed around the site. Triggering each at exactly the right moment was an engineering feat in itself—a far cry from repairing a CB on the fritz. The cameras were state-of-the-art machines worth more than Tim's annual salary, each capable of inhaling film at rates of anywhere from 5,000 to more than 200,000 frames per second. At that speed, the canister could empty in two seconds or less. His sequencer had to trigger each camera just seven-tenths of a second before the explosion. If activated too early, they would run out of film; too late and they wouldn't be up to speed. Tim knew he'd only get one shot.

The day of the test, he and Brown were hunkered down in a small bunker under roughly four feet of earth. They were less than a mile from the ANFO, closer than anyone else. At T-minus zero, observers saw a point of light, and from it the shock wave grew. It traveled upward and out, an expanding, translucent dome, the density of the air at its leading edge bending the light. They felt it in their feet before they felt it in the air. Tim and Brown were sure to keep their legs slightly bent; the sudden upward lurch of the earth would be painful to locked knees. A wall of sound washed over them and, after that, the negative phase, as a vacuum was created in the blast wave's wake. Tim felt every pulse as the fist of smoke, fire, and soil rose from a crater some 250 feet deep into the New Mexico sky. The test had gone flawlessly, exhilaratingly. It was a moment—among many others on the test range—that would stick with Tim: if even a nuclear explosion could be simulated, studied, picked apart, and known, what couldn't be?

The six young engineers at DRI became like brothers. When handling 30,000-joule lasers and military-grade explosives, each entrusted his life to the man working next to him. They traveled often, to White Sands or the Suffield Research Centre, near Medicine Hat, Alberta, Canada. The hours were long through the week, and on the weekend they were often together, grilling burgers and drinking beer. The tumbleweed ethos of the rootless, traveling geek was a lifestyle, and Tim became its embodiment: seldom home and usually without significant attachment.

By all appearances, he had been right about college. He had settled into a regular job—albeit a spectacular, adrenaline-charged one— where he was adored by his boss and colleagues and could be fully himself. Although his schedule now left little time for going up to Red Rocks and watching storms, his path was unfolding better than he could have hoped. He was a young man getting acquainted with life on his own—a happy, carefree, relatively unmomentous life—or so it certainly seemed for a time.

Then, on a winter day in 1980, he met the woman who would become his wife. Kathy Videtich had big, watchful blue eyes above high cheekbones, framed by permed brown hair. Formerly of DRI's travel office, where she made accommodations and booked flights, she had recently been transferred to the chemistry division when Tim strolled up to her desk to request a few thermometers for an oil-shale project. Kathy couldn't help but appraise the trim, dark-headed young man.

It wasn't simply that he was handsome and had a nice smile, though she would remember these things about him after he'd taken his leave. The detail that stuck out about him was that this small fellow had brought with him two full sandwiches for lunch. She would carry that odd first memory of him forevermore, the image of a young man who'd retained the metabolism of a thirteen-year-old boy—and the rabid curiosity as well.

Tim was smitten from the start, but he wouldn't see her again for more than a month. By then, Kathy had left DRI altogether. Yet, by chance, on January 31, 1981, Tim and a friend sauntered into the bar where Kathy was celebrating her new job as a legal secretary in a downtown law firm. Her girlfriend showed interest in Tim, but it was Kathy who absorbed his complete attention. Within an hour, their chemistry was undeniable. Tim thought her beautiful and down-to-earth. Kathy thought he was unlike any other twenty-three-year-old she'd ever met. He had goals, dreams; he was going somewhere. At the end of the night, she furtively slipped him her phone number.

The fledgling romance rapidly intensified, in spite of Tim's demanding travel schedule. That was, after all, one of the reasons Larry Brown had hired him. He was young, untethered, able to split town at any time. But when he was home, he and Kathy were inseparable. One night, while she was cooking dinner for him, he looked at her thoughtfully for a moment and announced, "You're the missing piece to the puzzle I've been trying to put together."

On March 26, after less than two months of courtship, he asked her to marry him.

That April, they bought a place of their own in Lakewood, just southwest of Denver. The outside was brown brick with white siding. Inside, there was a big basement for Tim's workshop. Two tall maples threw deep shade onto the front yard in summer. The wide street was quiet and lined with similar houses, which emptied young children onto the sidewalks, front yards, and driveways. Tim had grown up less than two miles away; in fact, his parents still lived nearby. Tim and

Kathy's children could eventually attend Alameda High School, as he had. It was the perfect place to start a family.

In December 1981, they were married in a small chapel at the University of Denver, with Larry Brown and the full DRI crew on-site to celebrate. From the start, Tim believed that if they waited for the right time to have children, the right time would never come. So, in April 1983, Amy was born. Jenny followed in December 1984. With Kathy staying home to raise the babies, money was tight. To keep up with the mortgage, Tim worked Saturdays and some nights at a radio-repair shop near Broadway and Eighth, as he had in high school. Given the nature of his job, he was also frequently away from the family. Six days a week he was working, though it never seemed to wear him down. When he was home, especially on Sundays, he was in full-swing dad mode, changing diapers and making up games for the girls. "He's always gone at a hundred miles per hour," Kathy says.

Their family of four was all Tim could ask for. Then, on April Fool's Day 1988, he returned home from the test range to some surprising news. Kathy declared that another Samaras was on the way. Tim wasn't convinced at first; they had taken every precaution. They asked the doctor to confirm the test, and when there was no longer any doubt, he shrugged: "Okay, I guess we're having three!"

Kathy went into premature labor that September and was placed on bed rest and medication. On October 31, the doctor took her off the drugs. Though the child's due date was December 3, she expected to go back into labor within days. The contractions came on Tim's birthday, November 12, and within hours the couple welcomed Paul Timothy—all six pounds, twelve ounces of him—into the world. Tim now had a little boy he could dress up as a foam tornado for Halloween, replete with cutout bolts of lightning.

————

It was around the time of Paul's birth that the old familiar urge returned. It had gripped Tim as a young man and had since been subsumed by the demands of adult life. Now, he began to look up at the sky again.

Kathy had long ago accepted that Tim would never be like other husbands. His work was unusual and potentially hazardous, and she knew there were some things he wasn't supposed to talk about. He could not, for example, go into much detail about the Patriot missiles he tested. He traveled often, and there was always some project pulling him away. But this new thing was perplexing. Why couldn't he take up golfing, bowling—hell, even model airplanes like his father? Why did he have to start driving off in search of what everyone else runs from?

What was he looking for out there? What did he expect to find?

CHAPTER THREE

THIS LOVE AFFAIR
WITH THE SKY

T HE URGE RETURNS the way it first began: with a tornado on television. "Tornado!," a 1985 episode of *NOVA* on PBS, shadows a
team of storm-chasing scientists. They race across the plains, hauling
with them a hardened weather instrument dubbed the Totable Tornado Observatory, or TOTO for short. Their goal: to deploy it into the
toughest environment imaginable, the heart of a tornado. These men
don't wait for the beast to come to them, they hunt it down—they *chase*.

This is an altogether novel idea for Tim. Nothing has gripped him
like this since he was a kid, when *The Wizard of Oz* rooted him in his
chair at his parents' dining table.

He wants do what these men do.

Because a smart man does not simply decide one day to take off
after a tornado-warned storm, Tim's forays into chasing emerge gradually. In the late 1980s, Tim is like the swimmer dipping his toe in to
test the water. Whenever the big black clouds form over Denver during
the late-spring afternoons, he starts to drive out to the edge of town to
watch, the way he did when he was in high school. He doesn't know
how to forecast. He can't tell the difference between a garden-variety
rainmaker and a dangerous thunderstorm. He cranes upward at the

underside of its gusting advance, watching the clouds flow and break like waves across the surface of the ocean.

Then one day, while his son is still in diapers, this isn't enough. Tim doesn't stop at the edge of town. He keeps driving, farther out—and the next time, farther still.

He begins to tackle tornadoes in the methodical way he does everything else: he studies them, figures out how they work, just as he did many years before with his mom's blender. For the first time in his life he enjoys going to class and sitting at a desk. In 1990 he enrolls in a six-week, forty-hour basic meteorology and storm-spotting course through SKYWARN, a program that trains chasers to become the National Weather Service's eyes and ears on the ground.

Tim learns all about the parts of the storm he has seen but couldn't name: that a storm's monstrous rotating cloud base is called a *mesocyclone*, that the tornado emerges from the *wall cloud*, which forms at the bottom of the thunderhead. He learns why Tornado Alley is such a powder keg, with two powerful currents of air—one from the west, dry and warm out of the Rockies; one from the southeast, moist and volatile out of the Gulf of Mexico—colliding here each spring. The atmospheric boundary where they meet is called the *dry line*, a north-south fissure from Texas to the Dakotas, which acts as a reliable factory for severe storms.

There's a sequence, Tim learns, to thunderhead formation. At the dry line, the air current out of the west will often sit atop the air from the Gulf, in what's called a *capping inversion* or a *cap*. The top layer then acts like the lid on a heating pot, blocking the lower-level air from rising until it has soaked up enough of the sun's energy to boil over. The cap can sometimes snuff out a brewing storm, or it can be the starting gun for the biggest behemoths of all. If the volatile Gulf air can break the cap, it races upward, sometimes clocking more than one hundred miles per hour. Thunderheads like mountains in the sky form in moments—all out of a remarkable alchemy of humidity, temperature, pressure, and flow. All out of ingredients that the diligent chaser can monitor and log.

Tim teaches himself rudimentary forecasting the same way he taught himself electrical engineering. Before a chase, he plots a weather map on paper, with an arrangement of dew points and surface observations, usually gleaned from a call to NWS Boulder. After that, it's a matter of identifying the time and the place where the sky will explode, like a mountain of ammonium nitrate. To be able to predict, drive out, and reap the reward of that first crash of thunder overhead: it's better than hearing any number of radios crackle back to life.

In the years before smartphones, radar apps, or even the Internet, SKYWARN is the only outlet through which a chaser on the move can access real-time weather updates—short of finding a pay phone. The group has installed a liaison in the Boulder weather service office, who relays the latest information. One must simply set the ham radio frequency to 146.94 kilohertz and tune in.

Before long, Tim is finding himself in the right place at the right time. He's racing those big black clouds, antennas swaying like reeds from the roof of the Datsun, accompanied by only the sound of Morse code bleating from his ham radio and the baritone wind over the Colorado plains. Weather service radar can detect tornadic circulation within the storm, but that doesn't always mean a tornado is on the ground. That's where SKYWARN comes in. The organization's trained spotters provide what's known as ground truth. Tim starts volunteering for the role, and he quickly develops a reputation as a spotter whose reports are reliable. He isn't one of those guys who cries "twister" when he sees stray tatters of cloud or dense rain shafts. No, he's got an eye for this.

In the early years, his excitement seems directly proportional to the amount of gadgetry in the Datsun. The stuff accumulates like mounds of coral. When there is no room left in 1994, Tim colonizes a used blue '91 Dodge Caravan he dubs Dryline Chaser. He mounts a crude satellite receiver that allows him to access some low-resolution radar feeds midchase. By adjusting the elevation and azimuth of the dish, he can pick up the Weather Channel from even the most lonesome corners of the state. Next, he defiles the interior of the minivan by sawing a hole into its dash to hold a nine-inch VGA monitor and 486 PC. Now he has at his fingertips an electronic DeLorme map with every nameless dirt road. More than likely, much of this gear is found lying around the shop or scavenged at the Aurora Repeater Association's annual Swapfest. Tim is, if nothing else, a bit of a cheapskate. He can cobble together working instruments out of what looks like junk. Over at DRI, they call it kludging, which means being resourceful and making an invention work with what you have at hand. But for all the half-improvised, battery-draining toys in the Dryline Chaser, he considers the CD player to be his most essential tool. Neil Young and Eric Clapton stay in heavy rotation.

On storm days, Tim is out the door a little after lunch, beating rush-hour traffic with Pat Porter, his brother-in-law, to catch thunderstorms inundating the southern and eastern suburbs. The Denver Convergence-Vorticity Zone—a place where the terrain often creates its own swirling winds—is a favorite hunting ground. Or they post up at Barbecue Point, a flat crossroads near Aurora, where a lightning strike once roasted a couple of cows.

After the initial shock, Kathy acclimates to all of Tim's carrying on about storms. He doesn't start blowing money they don't have to spend. He remains a good husband and an attentive father. As far as vices go, chasing clouds each year when storms roll around isn't all that bad. Yet she can't help noticing that this love affair with the sky is deepening. He's joining the ranks of a distinctly fanatical club.

The lengths to which some chasers will go to preserve their freedom during the plains storm season, lasting from April through June each year, is nothing short of astounding. One weather nut scheduled his wedding for the dead of winter expressly because it meant that his anniversary would never conflict with his chasing. Another refuses to accept the leash of any nine-to-five gig that would limit his ability to fully enjoy the sacred month of May. "You can have a nice, cushy job, but to me it'd be a nightmare not to be able to have an adventure," says chaser Dan Robinson. "I don't want the big house and picket fence."

Tim begins to schedule his vacations at the beginning of May through the end of June. A few years in, word of his abilities has already begun to reach beyond the insular world of chasers. His face ends up on a 1992 cover of the local alt weekly *Westword*, which contains profiles of him and a few other Denver storm geeks. "Some call it a hobby, some call it an addiction," Tim tells the reporter. "I think it's more of an obsession with me." He says he dreams of buying acreage out where the storms darken the plains of eastern Colorado and erecting an antenna farm for his ham radio.

The beast of Last Chance notwithstanding, by the midnineties Tim has outgrown the meager twisters of Colorado, and the casual day chases in his backyard. He starts roaming farther and farther afield, in pursuit of the "Panhandle magic" of the Lone Star State, and the storied monsters of Oklahoma—the stuff he's seen only on tape. It is as if he's fallen under the tornado's spell, which lures him away from home each spring and lies dormant in the fall.

By 1995, the Dryline Chaser is now recognizable in the empty northeastern tip of the Texas Panhandle and all the far-flung locales that have been calling him. On the afternoon of March 25, Tim and Porter are filming in Lipscomb County as the sky starts to boil in ways most people hope never to witness. The wind blasts Tim's back and he peers upward to see something burrowing through the bottom of the storm, fine tentacles sweeping and probing the air. Next comes a sign that even a newbie couldn't miss: that first welter of dust kicking up out of the prairie. "There it goes," he intones. Each time he's caught on tape, there's wonder in his voice, as though this is his first time.

Tim and Porter are soon racing down county roads. They cross the border into Oklahoma, the odometer maxed out at ninety, the mesquite fence posts blurring past, and the tornado falling away across a prairie that stretches to the limit of the eye.

The two can disappear this way into the interior of the continent for days or a week at a time. As the obsession deepens, Tim develops his own vernacular to describe what he's witnessing. He calls sheer vertical tornadoes *cigars*. The inflow stratus are *beavertails*. He talks about storms the way a collector might describe a Shelby Mustang. "Hard knuckles on the anvil," he enthuses. "Striations on the side. This is more than just a classic!"

And he *is* a collector, after a fashion—but of images and experiences, not things. The trophy is the right forecast. It's being able to say he was there—that in the vastness of the plains, he found the needle. The euphoric rush of pulling up just in time to see the cloud wisps gather and descend—it'll never get old.

If he's that lucky, he can also tend to more earthly concerns. Tim usually sells the footage to local television stations to cover his costs. At the end of the chase, he drops a line to his buddy Mike Nelson, the chief meteorologist over at Denver's NBC affiliate, KUSA. Often, Tim is calling in from Limon or some other town at the edge of the state: "I got some great stuff out in Kansas. Would you like it?"

And at 9:30 p.m., he shows up at the station's front door with a tape, just in time to turn it around for the ten o'clock news. If it's from Tim Samaras, Nelson knows it's good. Tim is one of the few guys who shows up with images of the thing in action, not the wrecked houses of its aftermath. Around the station, he has become something of a folk hero, this friendly dude who appears periodically bearing unbelievable video. Nelson occasionally hosts Q-and-A seminars on weather, and Tim is a regular guest. An introvert no more, when he talks about tornadoes, his voice comes alive.

He once sends Nelson a video intended to dispel any illusions the audience might hold about the glamorous life of a chaser. He has the camera rolling inside his motel room in western Kansas, panning over its modest furnishings. Then he zooms in on a framed picture of the actor Michael Landon, best known as the father in *Little House on the Prairie*. The sticker on the frame reads MICHAEL LANDON SLEPT HERE.

Tim's hand can be seen gently nudging the frame aside to reveal a gaping hole in the drywall.

Such is life in pursuit of tornadoes. The miles are long, the food is often gastrointestinal dynamite, and the lodging is humble. Tim wouldn't change anything.

———

When the spring and early summer have passed, and the sky quiets down, life regains familiar outlines in the Samaras household. Tim drops the kids off at school on his way to work. To Amy's profound embarrassment, the family minivan bristles with all manner of satellite receivers and lengthy antennae, each wobbling as the garish vehicle comes to a stop in front of her gawking classmates. While Kathy helps lead Amy's Girl Scout troop, Tim becomes the leader of Jenny's troop. He's believed to be the first male troop leader in the state. The little girls in his charge have mostly been raised by their mothers, and Tim's style of den-mothering is new to them. They don't knit or bake. Instead, they build model rockets.

Tim is also a popular attraction on Career Day, when Jenny proudly parades her father in front of her spellbound classmates, none of whom can claim a storm-chasing dad. He's equally effective when it comes to homework, especially with any assignment that might involve engineering. Tim is often far more invested in Jenny's science projects than she is, and the result—say, a tiny car powered by a mousetrap—always betrays a degree of technological sophistication she worries her teachers will find suspicious. Either that, she laughs, or they believe they've discovered a budding prodigy.

The childhoods of Amy, Jenny, and Paul fill up with uncommon sights. During one Fourth of July, Tim ignites a small chunk of solid propellant he'd scavenged from a Titan rocket. The kids watch from the safety of the living-room window as the fuel hisses and sends up a pale blue pillar of flame—a small taste of his work on the test range. As they get older, he also gives them glimpses of his all-consuming passion. They don't see much when he takes Amy chasing. But when he brings Paul and Jenny with him on a brief jaunt near Aurora, they find a funnel cloud hovering like a white proboscis over the foothills.

"Paul," Jenny's breathless voice can be heard saying in Tim's recording, "can you see it?"

Kathy has never chased with Tim and has no real desire to do so. What could possibly be so enjoyable about driving for days on end from dawn to dusk? She'll never find out, and she's fine with that. She has instead come to understand Tim's chasing as another odd wrinkle of the one-of-a-kind man she loves—a mostly harmless preoccupation whose biggest cost is Tim's absence once May and June arrive.

It is often said of Tim that to chase as frequently as he does, he must have an incredibly tolerant boss and, more important, a saintly wife. Chasing has been known to place an insupportable strain on even the sturdiest relationships. "I have had numerous girlfriends leave me because of storm chasing," says Ben McMillan, a friend of Tim's. "You're a normal person eight months out of the year, but then spring comes and your life completely changes." Infidelities have also been known to occur on the road, a testament to how the long weeks away can corrode.

But Tim and Kathy are a special pair. They both know how lucky they are to have found each other. When he's gone, Tim does all he can never to make her worry. Before he settles in for the night in Kansas or wherever, he calls to let her know that he is safe, and to tell her about his day. And she tells him about hers, sending along updates from their two beautiful girls and son. The connection holds strong, whether they're next to each other on the couch or separated by hundreds of miles of flatlands.

Kathy tries as best she can not to dwell on the inherent danger of her husband's pastime. She considers it, or course, but always returns to the same conclusion: This is Tim, the deliberate, conscientious man she can depend on. "I believed in him, and I knew him, and I guess I just had complete faith that he would stay safe," she says.

For now, there's little to fear. Tim has no reason to place himself in danger. Chasing is about tape, film, and experience—it's about testing himself against the inconceivable complexity of the atmosphere. But soon, he will need more—and he'll be able to do more. As anyone who knows him would tell you, Tim has never been content merely to observe. Ever since he was a boy, his need to understand has been compulsive, consuming. Tim Samaras can't help but take something apart just to figure out what makes it tick. And over the last few years, he has built up a skill set that could allow him not simply to acquaint himself with the field of tornado science, but to drive it forward.

CHAPTER FOUR

THE SPARK

A S A STORM assembles, as the clouds darken and swirl, scientists are often blind to what sets a serious thunderstorm apart from an extraordinary tornado. It could be as subtle as a shift in the direction of the wind, or a break in the clouds that lets the sun shine through. The direct causes are rarely recognized as they happen. The triggers of an EF5 may become clear only after the sky has exploded.

In the spring of 1997, Tim receives an unexpected phone call. As it rings, he has no idea that this conversation could change the course of his life, or that he sits amid a vast sea of forces just one nudge away from alignment. He'd never be able to know it at the time—but the click, as he picks up his phone, is the shift that sparks the storm.

The man on the other end of the line is a mechanical engineer from Huntsville, Alabama, named Frank Tatom. He runs his own firm specializing in the dynamics of turbulent flow and explosive damage, and he has found Tim's name in a search for tornado chasers across the Midwest. He has a proposal that he thinks Tim will want to hear.

Eight years ago, Tatom begins, a violent twister entered the southern sector of Huntsville before the sirens could sound. Nobody saw it coming. The tornado cut across highways choked with rush-hour traffic, killing eleven people. Another eight died in their homes and businesses. Two other victims held on for a while before succumbing to their injuries in the hospital. Huntsville looked like a war zone. The damage path was a half mile at its widest, and nearly nineteen miles long.

Tatom had been reeling just like everyone else, but amid all the heartbreak and pain, one story struck an odd note to his engineer's ear, if only for its inexplicability. A man working at a service station told Tatom that minutes before the tornado arrived, he had the hood up on a woman's car. Suddenly, the ground beneath his feet began to tremble. Even the motor in front of him seemed to vibrate like a tuning fork. He didn't know what was coming, but he grabbed the woman and pulled her into the service station. "The man took her to his office, threw her down near the counter, and jumped on top of her," Tatom relates to Tim. Then, they heard the roar, faint at first but soon deafening. The building took a direct hit, and the two were flung into the street, rattled but miraculously alive.

The story made Tatom's hair stand on end, and the more he looked around, the more he discovered the man's experience wasn't unique. The choir leader at Trinity Methodist felt the vibrations through his feet as he sheltered in the church basement. Both a police officer and the city's emergency manager reported the same sensations. Tatom went to the nearest US Geological Survey office. The geologists there confirmed a low-frequency seismic signature—a series of apparent shock waves that accompanied the storm.

In the years following the tornado, Tatom couldn't stop thinking about the vibrations on that terrible day. Could the wind really have caused the earth to tremble? He and an expert in soil dynamics and seismic signals at the University of Alabama decided to chase down the hypothesis. With their combined expertise, in 1993, Tatom and Dr. Stanley Vitton concluded that the tales of tremors weren't at all far-fetched. In fact, they surmised, the most violent tornadoes transferred a jaw-dropping amount of force into the earth—the energy equivalent of half a ton of TNT per second.

Tatom's bold idea finally clicked into place: What if seismic sensors could have warned Huntsville back in 1989?

Tatom has since devised a prototype instrument package he calls the snail. Its components are quite simple: an aluminum basin from Walmart houses a battery, and a recorder microprocessor that's connected by cable to a geophone—a sensor used to detect seismic signals with three sensitive transducer prongs plunged into the ground. Theoretically, the device should be able to detect tornadoes at a distance of twenty-five miles. The idea is to position snails at intervals around the city as a kind of early-warning network.

By the time he calls up Tim, NOAA and the Department of Commerce have already awarded Tatom a Small Business Innovation Research grant to take the snails to Phase I: production and testing. He now needs to prove his concept by deploying the snail near a violent tornado. This is where Tim comes in. Tatom is just an engineer, he explains—a man of equations and microprocessors and laboratories. He knows nothing whatsoever about hunting twisters. But Tim's prowess as a chaser precedes him.

Can you get my invention close to a tornado? Tatom asks. *Can you help me find out if it actually works?*

What he's proposing sounds a great deal like the *NOVA* documentary Tim saw years ago, in which scientists lugged their high-tech TOTO probe across the plains in pursuit of scientific knowledge. It's the very notion that started Tim chasing.

Send the snail, Tim replies.

Before the spring storm season arrives, a single snail—number five in the fleet—is delivered to his doorstep in Lakewood. Examining the instrument, fashioned out of household fixtures and seismic sensors, Tim may as well be looking into his future. If the phone call was the spark, this glorified sink basin is the first rising thunderhead. He practices deploying the snail and its geophones in his yard, closely watching the weather patterns for the shape of the year's first tornado. He won't wait long.

Tim and his brother-in-law, Pat Porter, gear up to hit the plains on Memorial Day weekend. Tim is more excited than he has ever been about a chase; this one will be unlike any he has ever embarked on. Gathering footage and sending ham radio updates to SKYWARN have never required proximity. Deploying the snail, however, will mean heaving aside nearly every piece of chaser wisdom he's ever absorbed. What happens to the first and only rule—don't get too close—when close is exactly what Tatom needs?

Tim has never considered himself a daredevil. He has always avoided entering the rain and hail core that directly surround the funnel. He stays out of the tornado's way. But if he is to succeed now, he may well have to deliberately enter the path. Tim has enjoyed storms for roughly a decade now, but there's something newly electrifying about working for Tatom. It's not just the adrenaline rush, though that's undeniable. What strikes Tim now is a new sense of purpose. It is time, he decides, to use his storm-tracking skills for more consequential ends.

———

On May 25, 1997, Tim and Porter drive through northwest Oklahoma, near the city of Woodward. To the north, storm cells are popping up. They cross into Kansas and soon park next to a wheat field ten miles north of the border. Just south of Rome, Kansas, at a dogleg in the Arkansas River, their necks crane to the west, across cottonwood-lined creeks and tracts of wheat and sorghum, toward the bluff crags of a thunderstorm. They can't see it yet, but the NOAA weather radio indicates that somewhere behind the far line of trees, beneath all that rain-sheathed darkness, a tornado moves east-northeast. Tim centers his camera frame and begins to narrate: "There's a reported tornado on the ground *right now*."

From this distance, he can see the broad span of the storm, from its base to its mushrooming anvil. "A classic," he exclaims. "We have *monster* rotation here. Likely, we're going to have to move east. Just a little bit, that is. Time: 7:32. This is an incredibly impressive storm."

As it nears, a form emerges out of the murk in the west. It's like the trunk of a great tree, the color of a livid bruise: hundreds of yards across, its canopy fans into restless clouds. With the inflow at his back, Tim can feel it breathe. He watches the wind thresh the wheat at the edge of the road and sees a cold blue glow spill through the clouds above the tornado. As windrows of dust kick up above the tree line, the tornado undergoes a sudden and detectable hardening. It's feeding, growing stronger. "No question about it," Tim announces. "That is a huge, *huge* tornado."

He stands there for a while longer, projecting its heading with his eyes. Then he piles into the Dryline Chaser and, for the first time, drives toward its path. He and Porter reposition onto a gravel road adjoining an unfenced, fallow field, and when he's satisfied, Tim stops and steps out into the wind. The dark shag on his head tosses in the rising gale. He wears blue jeans, a white T-shirt, an unbuttoned denim jacket, and running shoes. He opens the minivan's rear hatch, revealing a compartment bursting with luggage. After retrieving a large, stiffened case, he sets it gently down onto the road, pops the latches, and uncovers the humped profile of the snail, which Tatom has painted bright orange for visibility. With the snail tucked securely under his left arm, Tim sets out.

This time, he isn't following at a cautious distance. The tornado is coming for him, and instead of running away, Tim is jogging in the direction of the darkness. It lurks just across the field, beyond the windbreak trees, as Tim jumps off the road and onto bare dirt.

He kneels, lowers the shell, and drives the geophone's prongs into the silt loam. Lightning stitches the horizon, and there he is, crouching in the furrows as this shadow like something out of a fable rumbles and moans. He points toward the tornado with his left arm and angles his right arm to the northeast, gauging trajectory. He says something, but whatever it is gets lost in the wind and thunder. Tim rises and hustles back to the minivan. There's a change in his voice when he speaks again; it's never sounded like this on a chase before. Usually he's full of glee and boyish astonishment. Now his voice sounds forceful, deadly serious, that of a man at dire work: "Time: 7:45. Snail deployed."

They wait for a moment until dust hangs like a curtain in the air and Tim knows that the big winds are on their way. They bail ten miles to the southwest and to safety, parking at a Motel 6 near Interstate 35, in South Haven. The power is out and the motel and its parking lot are dark. To the north, electrical transformers arc briefly like lightning bugs in the night.

The next morning, Tim and Porter return to the deployment site near Rome. Trees with nude branches lean against a nearby house. Power poles are strewn across the field like felled timber. Tim calls Frank Tatom with the news. He's talking fast. The snail was close, and the instrument survived. More important, it's still recording.

In fact, of all the snails in the fleet Tatom has entrusted to various chasers, Tim is the only one who gets near a tornado. He's good at this. What's more, he has flourished in the excitement, the danger, and the purpose. This can't be the end, Tim thinks. He has gotten a taste of what it means to become something more than just a watcher, and Tatom has shown him the way.

The federal government ultimately declines to fund a full seismic tornado-detection network, concluding that radar's early-warning capabilities are superior. Nevertheless, Tim can't help notice that Tatom had enticed the government to pay for the development of his dream instrument. What is stopping Tim from creating one of his own—a different device capable of prying loose the secrets his inquisitive mind has sought over countless miles of plains highway?

Tim knows that for all the work that went into the landmark TOTO project a decade before—and for all the people it inspired, from Tim himself to the scriptwriters behind *Twister*—the storm scientists in charge never actually succeeded in getting a direct hit with their probe. They failed to pierce the tornado core, and no one has succeeded since. Their dream of seeing into the heart of the tornado was never realized.

In Tim's mind, if their failure has proven anything, it is that the field could use an engineer's touch. The TOTO probe looked like an oil drum and weighed about as much as a full one. The snail's largest component is an actual aluminum *sink basin*.

Field science deserves a better probe. Tim is now awakened to the possibility that he might be the guy who can build it. More than that, he thinks he could be the one to get it inside a tornado. It's the kind of pipe dream that your average chaser might alight on in a moment of inspiration and then let smolder and fade.

That's not Tim, though. The laboratory at work is brimming with the right technology. He has already learned how to measure the blast wave of an explosion. What's so different about the wind? In fact, he realizes he's been training to build just such a device since he was a little boy, hunched over old radios in his bedroom. The question isn't "Am I capable?" The question is "What should I build?"

The answer arrives in 1998 in the form of an announcement in *Commerce Business Daily*, a clearinghouse for government contracts. Tim's eyes skim over the page in the course of his work. The National Oceanic and Atmospheric Administration is seeking proposals for the development of a hardened instrument capable of obtaining measurements from *inside* tornadoes. This is it. With just a few column inches of print, NOAA has handed Tim the very pursuit he was looking for. It's the project of his lifetime. The scientists who wrote this notice don't know it yet, but Tim Samaras of Lakewood, Colorado—a guy who, in his words, "blows shit up for a living"—is the chaser-engineer they seek.

CHAPTER FIVE

CATCHING THE TORNADO

B Y TAKING UP a task that seems like such a serendipitously good fit for a man with Tim's background, he's in fact taking up a mantle with decades of portentous history. If anything, building a device that can survive the obliterating tornado core—much less placing it there—is like walking up to Excalibur, plunged deep in stone, and giving it a hearty tug. Many have tried. All have failed.

If he knew in advance just how difficult it would be to take on the Black Wind—and how impossible it has been for his predecessors—he might think twice. To chase and watch is one thing. If you seek no answers, you court little danger. But prying open the tornado's secrets is an altogether different endeavor.

Every fleck of insight science has gained into massive thunderstorms and tornadoes has been hard-won. Even seventy-five years ago, during the Second World War, as humanity was unlocking the inner workings of the atom, tornadoes were still considered nigh unknowable. That people like Tim can now, with a little luck, occasionally predict, track, and chase them is a testament to the line of tenacious researchers who refused to look at the storm and accept it as inscrutable. At each stage, researchers have had to chip at the limits of our understanding one basic question at a time—*What is a thunderstorm? When can a tornado form? Can we ever hope to predict where it will strike?* Every answer has revealed a more astonishing, more complex presence on the other side than its investigators could have expected. The tornado has proven a foe as worthy and wily as any fabled monster.

Now Tim hopes to join the lineage of those who've stolen away with its secrets. For all he knows—about chasing, about TOTO's failings, about how to quantify extreme force—he still can't make out the full shape of the undertaking before him. To grasp the sheer enormity of the quest he has chosen, and the foe he will face, requires understanding Tim's predecessors, and the story of their fight to lure tornadoes out into the realm of the known.

———

Well into the twentieth century, the tornado still resided more firmly in the world of myth than reality. Towns in Tornado Alley seldom saw the vortex coming until it was already at the door, and survivors rarely even had the language necessary to describe the malign force they'd witnessed. Those who emerged from the worst modern tornado disaster, a 1925 monster that left a three-state trail of destruction, could only report that the culprit was a "smoky fog" or a fell "blackness" that had descended upon their towns.

Until the early 1950s, official policy forbade even the utterance of the word *tornado* in weather forecasts. The government was convinced that citizens were no more sensible than stampeding cattle, that entire cities would descend into hysteria upon hearing the dread word, resulting in far more fatalities than the thing itself. *Tornado* was a word of power: deadly if spoken; deadly if left unspoken. Better to leave it unsaid, decided the US Army Signal Services, and later the US Weather Bureau, since few within their ranks believed tornadoes were actually predictable. The forces causing the winds to coalesce were shrugged off as the acts of a jealous God. All that meteorologists could do was catalog their epidemiological particulars: deaths, injuries, property damaged.

It took two freak storms—at Tinker Air Force Base near Oklahoma City, both striking within days of each other in 1948—to shake the institutional opposition to tornado forecasts. The first was an utter shock, unforeseen, smashing thirty-two of the Air Force's most advanced aircraft at a cost of more than $100 million in today's dollars. Five days later, two weather officers noticed that the morning's atmospheric conditions bore an uncanny resemblance to those presaging the storm that had just ravaged the base. Despite the one-in-a-billion odds of two tornadoes striking the same spot within a week, the officers issued the first-ever operational tornado watch. They would be on the chopping block if proven wrong. But then, just after 6:00 p.m., a "yellowish" vortex like a giant "radish" dropped down and swept across the runways. The officers' prediction saved Tinker millions of dollars in salvaged military hardware, and it offered the first glimpse of a pattern.

Even so, progress was sluggish. The first civilian tornado "bulletins" wouldn't be issued until four years later, in 1952, and few subsequent forecasts would speak of tornadoes. Why risk career suicide by attempting to predict such a fickle event? Those warnings that were issued were often inaccurate or unreliable, just as meteorologists had long feared. The areas encompassed by any given watch were so vast as to be nearly useless—a parallelogram- or trapezoid-shaped bulletin zone of some 38,000 square miles. Forecasters simply did not know enough about how tornadoes formed to issue any more specific or accurate warnings.

In 1951, the Weather Bureau finally set out to solve the problem. They launched the first coordinated effort to understand the cause of tornadoes, the aptly named Tornado Project, which would run through 1953. The bureau constructed a network of instrument stations, called a mesonet, throughout Kansas and Oklahoma, attempting to log the atmospheric conditions preceding and surrounding tornadoes. But over the course of the project, researchers learned precious little about the nature of the tornado. Twisters seemed to be utterly repelled by the network. Even trained scientists couldn't resist imbuing the storms with anthropomorphic caprice. "Unfortunately for meteo-rological knowledge," the project's leader drily joked, "the setting up of the Tornado Project system seems to have provided the people of Kansas with the best tornado insurance they ever had."

This first effort set forth a theme that would repeat time and again in the history of tornado research. Tornadoes were the essence of ephem-era, often so brief and random that it was as if they conspired to keep their own secrets. From the 1950s to the present, those who have tilted headlong after the vortex itself have, more often than not, caught noth-ing but air. Fate has instead rewarded those who have been more circum-spect, targeting the larger storm or paying no mind to the vortex at all.

Tornado science has proceeded from broad to small, from com-mon to rare. First the thunderstorm was charted and explored; next came the supercell, the specific variety of storm that tends to spin off twisters; only once those two were plumbed did the vortex begin to shed its mystique. From stratosphere to surface level, science would march down the scale. Only at its lowest levels—where Tim has set his sights—does it continue to fend off all comers.

———

Every tornado is spawned by a thunderstorm, those mammoth atmo-spheric engines that roar and spark and race across the plains, occa-sionally channeling their full wrath down upon some tiny hamlet in the sea of grain. By deciphering the puzzle of the thunderstorm, re-searchers would gain their first important toehold toward understand-ing the tornado.

Launched in 1945, the Thunderstorm Project, a sister program to the ill-fated Tornado Project, set out to solve another freak weather event that was tormenting the postwar nation. The nascent commer-cial air industry had just begun to boom, but as it rose, its planes were falling from the sky at astonishing rates. The new craft of choice, the Douglas DC-3, was crashing regularly amid the extreme turbulence of thunderstorms—a rising toll that included one grim disaster that killed a sitting US senator. When Congress called upon the Weather Bureau to stem the crisis, it turned to the esteemed meteorologist Dr. Horace Byers, out of the University of Chicago. And what Byers ultimately discovered would extend well beyond air travel.

Byers's ambition was nothing short of mapping the wind. Taking advantage of the nation's postwar surfeit of pilots and aircraft, he assembled his own fleet of ten Northrup P-61Cs, known as Black Widows. He then retrofitted each plane with sensors to track their encounters with violent air currents, and he prepared them to penetrate storms at five elevations, from 5,000 to 25,000 feet. Stiff-winged "night fighters," designed for navigating by radar in the pitch dark, the P-61Cs were about as rugged as a plane could get in 1945—and Byers put them through the wringer.

The fleet intercepted every malicious thunderhead that formed near the bases of operation in Orlando, Florida, and Wilmington, Ohio—no matter how ominous or how riven with lightning. Through the storm seasons of 1946 and 1947, the pilots encountered seventy-six thunderstorms and conducted nearly 1,400 penetrations. The aircraft were struck by lightning and pummeled by hailstones that left three-inch indentations in metal nose cones and cowlings. They were buffeted by extreme updrafts and downdrafts, one of which shoved a P-61C a vertiginous 500 feet earthward in seconds. Yet for all the stomach-pitching turbulence, not a single pilot was lost.

Through it all, Byers and his team did exactly what he'd hoped—they mapped the storm. To his surprise, he discovered that a thunderstorm was not a solitary, massive edifice, but rather an agglomeration of cells, complete with complex structures and life cycles. It was as if he'd stumbled upon an alien life-form. Up in the clouds—*made of clouds*—was a being that could grow and divide, feed and die. In his analysis, Byers was able to lay out a model. He charted the set of ethereal ingredients that come together and dissipate with every storm system. And in doing so, he offered science its first step toward being able to know storms and tornadoes as anything but the hand of God.

Each storm, Byers discovered, results from a confluence of moisture, heat, and lift. Warm air near the earth's surface is unstable: it
wants to expand and to rise. As it ascends, it creates a vertical current
called an updraft—which can be boosted by the right winds. The updraft builds over time, conducting a stream of warm air into the cooler,
upper layers of the atmosphere. Here it inevitably starts to shed its
heat, which causes the updraft's moisture to condense and create a
cloud. One can often see cumulus towers billowing up to the ceiling
of the sky just before the storm strikes. If the updraft is strong and the
air especially warm and moist, the clouds will keep on rising, higher
and higher. But once the force of the updraft can no longer support
the weight of the condensing water droplets—what goes up must come
down.

Byers found that a storm cell is like a lung. The updraft is the inha-
lation. Now comes the blow. All that moisture hurtles down to the surface in the form of rain. Along the way, the rain cools its surroundings,
creating a dense flood of descending wind, known as a downdraft. This
is the chill you feel at the passage of a storm. This, Byers found, is the
windy sledgehammer that was pounding DC-3s out of the sky.

Within a short time, the frigid outflow marks the beginning of the
end of the storm. Lungs can't inhale and exhale simultaneously. Without the energy source coming from the updraft, the storm devolves
into a downpour of wind and rain and soon expires. Byers summed
up the progression in a reliable three-stage life cycle: the storm builds
slowly as warm air rises; it reaches maturity once the condensation gets
too heavy and begins to fall; and it dies once the outflow drowns the
updraft.

For the practical applications of the Thunderstorm Project, Byers
was able to prove what Congress had hoped—that a trained pilot and
an aircraft equipped with radar could detect and avoid the most treach-
erous drafts of a storm. Thanks to these intrepid test pilots, death while
traveling by commercial air has become most unlikely. But for tornado
science, what Byers and his cohort missed was the next leap forward.

Within his own data and radar images was a notable exception to Byers's final rule. There was one rare type of storm that didn't quickly drown itself out; it could live and feed for hours, traveling far, growing immense. And it was this distinct breed that was the mother of the twister.

Keith Browning—an English scientist working at the Air Force Cambridge Research Laboratory in Bedford, Massachusetts—was the man who finally discovered and named the supercell. Picking up where Byers and his crew had left off, he succeeded in identifying the essential ingredients of this rare class of storm, allowing us to begin to differentiate the average rainmaker from its destructive doppelgänger.

Browning's breakthrough came while he was studying radar imagery from a nasty 1961 tornado near Geary, Oklahoma. He noticed several peculiar attributes of the parent storm. For starters, while a line of storm cells was drifting regularly northeast, the one tornadic cell had cleaved away and was moving obstinately due east. As the other storms unleashed short-lived deluges and promptly dissipated, the lone cell continued to thrive for hours. It was also far larger than the others, standing eye to eye with the cruising altitude of a modern jetliner. More perplexingly, it exhibited a curious structure—a "vault" appeared on radar, a precipitation-free zone soaring high into the heart of the storm. What could possibly explain a persistently dry region within a proven rainmaker?

Browning found more and more storms with the same features, and he soon developed a theory: The longevity of these storms owed to an updraft on steroids. It was carving out a space in the center of the storm with velocities so intense that rain and hail simply could not fall there. This wasn't the garden-variety buoyant column of air that would soon drown in the smothering downdraft. Browning's updraft existed in a state of continuous reconstitution, propagating forward along a stream of buoyant, energy-suffused air like a wave on the surface of the ocean. So long as the energy source remained unobstructed, this "supercell" could conceivably continue to thrive like a perpetual-motion machine.

What made the updraft in Browning's model work was its unique ability to rotate, a quality that Byers had already noticed in some of the strongest storms. The updraft owed its freakish strength and longevity to a confluence of spiraling winds that continuously funneled warm air all the way up to the storm's highest levels, like a fuel injector.

The supercell thunderstorm was a complex organism that required an unlikely combination of interdependent elements brought into just the right alignment. As Browning and others soon realized, nowhere else on earth do these elements seem to come together as in the North American plains. It isn't just the hot, volatile air or the pressure cooker of the "cap" that turn the Great Plains into Tornado Alley. In the turbulent springtime months, the converging wind currents are coming from different directions, at different elevations, and at different speeds—which makes them prone to generating wind shear and the crucial rotation needed for tornadic storms. With the right winds, Browning saw, supercells pop up all along the dry line. When winds out of the southeast, the southwest, and the west—all at different levels— hit a rising column of warm air, they spin it like a top. The rotating updraft that results, what would become known as a mesocyclone, is a staple of Tornado Alley and the engine of a supercell.

As a final piece, when the mesocyclone is met by the jet stream, some six miles above the earth, the storm is able to reach its full strength. The powerful upper-level winds, in concert with the tremendous velocities of the mesocyclone, shunt the precipitation away from the throat of the storm, thereby preventing the suffocation that inevitably shortens the lives of common thunderheads. The mesocyclone and jet stream are what allow Browning's storm vault to form.

When wind shear, the cap, a powerful jet stream, and volatile quantities of atmospheric instability pile up over the Great Plains, batten down the hatches—the titan of the sky is coming. Unlike its disorganized counterpart, the isolated supercell thunderstorm can swell to immense proportions, measuring miles across. It may contain straight-line winds capable of toppling telephone poles. Its updraft, a screaming hundred-mile-per-hour vertical vent into the atmosphere, can loft and suspend rain particles. These can then freeze and weld together into grapefruit-size hailstones, plummeting to the earth at terminal velocities. And when the rotating updraft reaches all the way down to the surface, the supercell storm has the power to summon forth the most destructive phenomenon of all, the Black Wind.

Browning was able to narrow the field of view, surmising that the vast majority of tornadoes, especially violent ones, emerged from supercells. But in his time, the tools didn't yet exist to limn the difference between a nontornadic supercell, and the type whose rotating winds will reach the surface.

To that end, tornado science's biggest breakthrough would come with the arrival of Doppler radar in the 1970s. Since 1953, it had been known that tornadic storms create a unique image on radar called a hook echo—so named because the radar picks up the arc of rain or hail that wraps around a strong mesocyclone. But before the arrival of Doppler radar—an advance that allowed radar to track movement in real time—meteorologists were unable to learn much about the supercell's evolution or internal structure.

The first Doppler scan of a tornado would leave little doubt about just how much the technology had to offer. Like most advances in the field, it came about only with a rare alignment of chance elements. Throughout the sixties and early seventies, a small number of Doppler weather radars had started dotting the states, but it took until May 24, 1973, for a recordable tornado to form and touch down within range of any. That day, researchers at the National Severe Storms Laboratory in Norman, Oklahoma, noticed the first signs of a broad vortex forming several miles off the ground, and just within range of their Doppler. They scrambled to record its every moment—as the circulation focused, touched down, and marched six miles through the open country into nearby Union City. Their scan, of an entire tornado from birth to death, was historic—but the fruits of their labor took several more months to reveal themselves.

In the aftermath, researchers Rodger Brown, Donald Burgess, and Les Lemmon at the NSSL pored over every angle, every azimuth, on the magnetic tapes that recorded the day's radar feed. When they struck upon a place where wind shear and velocities spiked over a short distance, they assumed they had found an error in the data. At that moment in the storm's evolution, there had been no tornado on the ground; the increase in wind speed should have been smooth and gradual. Yet even when they checked different elevations, the same steep velocity gradient persisted. The researchers consulted the photographs taken from the field at the same time and identified a small funnel protruding from the cloud base, not yet in contact with the ground. In fact, it would be another forty-one minutes until touchdown.

What they'd stumbled across was no error. Their device had identified the first embryonic stages of a tornadic storm. In the race to warn those in the path, this was monumental. It meant that Doppler could reveal areas of concentrated rotation a few kilometers off the ground, a pattern detectable tens of minutes before touchdown. In the coming years, Doppler and follow-on advances would spread like wildfire across the meteorological community. While not all tornadoes produced visible radar signatures in their infancy—and not all signatures resulted in tornadoes on the ground—forecasters finally had *some* way of peering into the storm and spying a brewing tornado prior to its arrival. In the long history of tornado science, researchers could finally boast a clear victory over the vortex.

————

Through the thunderstorm, to the supercell, to the first spiraling wisps of the nascent twister, scientists had chipped away, steadily earning their insights. Yet as the field tried to isolate and discern the features of the funnel itself—What made the vortex descend? What determined its strength, or how long it would churn?—the tide suddenly turned.

Around 1980, after decades of scientific progress, the tornado seemed to set in its heels. At the lowest elevations—what's referred to as the boundary layer—the twister refused scientists any further victories.

Tornadoes were such an improbability in the first place, such a miraculous confluence of variables that the trigger might be as inscrutable as chaos theory, and the beating of a butterfly's wing. Vastly more data would be needed to detect any pattern. Doppler had proven its ability to capture in unprecedented detail the life cycle of a storm—but it had serious shortcomings. It struggled to capture the near-surface parts of the storm that scientists most prized. Unless a radar installation was practically run over, trees, houses, even the curvature of the earth, would conceal the vortex—to the point where scientists weren't able to discern whether the wind was stronger near the surface or farther up. The boundary layer proved too low for the beam to reach.

It did not help that the technology was wholly reliant on chance encounters. Any particular spot in the country will meet a tornado, on average, once every 4,000 years. The odds were disconcertingly long that one would land within close range of a set of stationary meteorological sensors. By stringing mesonet instrument stations across the Oklahoma prairie through the sixties and seventies, the National Severe Storms Laboratory was able to improve the long odds. And the right storm in 1973 did yield its groundbreaking Doppler data set. But the run-ins were painfully rare, and storms that hit could be even more agonizing than those that missed: In 1977, when a tornado finally touched down near a site in Fort Cobb, it got *too* close. The site's sensors were not designed to withstand tornadic wind speeds, and no useful data could be gathered.

Finally, there were a vast number of questions that Doppler would *never* be able to answer. The technology could say nothing about the temperature, humidity, or pressure inside the tornado. These are the fundamental data points of meteorology—what scientists consider essential to knowing where the wind will go, where it comes from, and what exactly is driving it. No fine-grain predictions of tornadic behavior would be possible without them.

Yet even as Tim Samaras takes SKYWARN classes and trawls Last Chance, these same basic questions remain unanswered. Upon reaching the unknowns at the tornado's lowest levels, the science stumbled and sputtered.

A new tool, the next breakthrough, was needed.

In 1979, Dr. Al Bedard, a scientist at NOAA's Wave Propagation Lab in Boulder, Colorado, knew it was time to risk a more daring approach. A stationary weather instrument wasn't ever going to glean enough data. So while attending an after-hours party during a NOAA conference, Bedard approached Dr. Howie Bluestein, a member of NSSL's Tornado Intercept Project, with a long-shot idea he'd been considering. Bedard had been developing hardened instruments for taking atmospheric measurements at airports, instruments that could hold up to a direct hit. But what if, he asked, he were to incorporate some of these same instruments into a mobile package instead? Such an "in situ" probe wouldn't have to wait for the tornado or fight the 4,000-year odds. Imagine a portable twister laboratory engineered specifically for the chase. If he built such a thing, Bedard wanted to know, would Bluestein be interested in deploying it?

Bedard didn't need to belabor the importance of such a new tool. It might have been half-raving, but Bluestein signed on to the endeavor. So the next evolutionary leap in tornado research—or perhaps its most frustrating period of all—got its start over cocktails.

By 1980, Bedard had constructed the instrument package: it was an ungainly cylinder roughly the size of an oil drum, made of half-inch-thick aluminum plate. The Totable Tornado Observatory would house sensors for measuring pressure, temperature, humidity, and wind speed, hardened enough to withstand velocities exceeding two hundred miles per hour. Because deploying in the path of a tornado was bound to be a frightful experience—and Bedard did not intend to deploy TOTO himself—he sought to make the process as foolproof as possible. Lying horizontally in the back of a pickup, the sensors would remain inactive. But as soon as TOTO was upright and winched down a ramp into place, the battery would kick in and its sensors would start recording.

In preliminary field trials in the summer of 1980, Bluestein tweaked the hardware and improved on Bedard's deployment technique, until he could unload TOTO in roughly a minute. Then, in 1981, scientists at NSSL and the University of Oklahoma started chasing storms. They used a government van with TOTO's bulky frame strapped down in the back. In an era predating cellular networks, Bluestein fed quarters into pay phones for storm updates from NSSL. When it was showtime, he chased by sight and chose his deployments carefully. They only had one chance, one probe, after all; and TOTO required relatively firm, level surfaces. Mindful of potential sources of debris that could strike the instrument, Bluestein was also conscious to avoid barns and trees.

For three consecutive seasons he chased, with success always just out of reach. In 1982, he was racing to get ahead of a fast-moving storm when he realized he was on a collision course. As he began to turn back, the tornado roared into view, snatching telephone poles out of the ground and flaying a nearby mobile home. He'd gotten close, but too damned close to deploy. During one infuriating chase, he pursued a tornado sitting over Altus Air Force Base in southern Oklahoma. Before the team could get into position, the tornado lifted without warning. This wasn't that unusual; the storm seemed to be producing twisters cyclically, so Bluestein continued ahead to where the next was likely to drop. *We'll get this one this time,* he told himself. But instead of the usual southwest-to-northeast track, the new tornado darted off to the northwest, eluding TOTO. "It was teasing us," Bluestein laughed. "Catch me if you can!"

By 1983, Bluestein had begun to despair. Short on hope and long on frustration, his group had TOTO tested in a wind tunnel at Texas A&M to calibrate its sensors and to determine its tip-over threshold. The results weren't terribly encouraging. Unless anchored or widened, a hundred-mile-per-hour gale would upend the top-heavy cylinder— the kind of wind you'd see in even a moderate tornado. By the end of the season, Bluestein turned TOTO over to other scientists at NSSL. After years of close calls and frustrating misses he was finished with this dangerous game.

For all the long miles logged, the device got tantalizingly close only once. Near Ardmore, Oklahoma, the following year, the NSSL team deployed TOTO on the outer edge of a weak tornado. The instrument promptly pitched over onto its side, damaging the sensors.

In every sense, Bluestein felt as though they had been tilting at windmills. The Totable Tornado Observatory was retired for good in 1987. "I said, 'I give up. Forget it,'" Bluestein recalls. "There are easier ways to do this.'"

The way he chose was mobile radar, which had become increasingly viable and cost-effective in recent years. The upside was obvious: One could still get around the 4,000-year odds, but the intercept was no longer an all-or-nothing proposition. Bluestein could keep his distance and position a mobile dish anywhere near the tornado, rather than only in its path. It wasn't a replacement for TOTO so much as another path altogether. "It would be fascinating to actually get inside the tornado and take a look around," he said. "Since we can't, we try to get close enough to aim our portable radar unit and measure the wind field in and around the tornado." Later, after Dr. Josh Wurman proved that the larger, more powerful Doppler radar could be adapted to the road, Bluestein mounted a heftier dish to the back of a passenger van. Where stationary Doppler would scan from dozens of miles away, Bluestein could haul his antenna to within a mile of the storm. With it, he achieved one of the first major breakthroughs in years. F5 velocities had previously been inferred indirectly through photographic analysis and damage assessments. His mobile radar, hauled into range of the funnel, proved that tornadic winds were far stronger than scientists had previously thought.

As mobile radar proliferated, a string of new findings followed, from Bluestein, Wurman, and others. The new technology triggered a badly needed fertile period for the battered and bruised researchers.

But the radar's beam still couldn't reach the place TOTO was designed to go, where the tornado and our world meet.

Even as the twenty-first century neared, there were no accepted theories for how tornadoes formed, for how fast the winds inside could blow, or for how far the pressure could fall inside the core. To demystify these questions, one further outfit, a multimillion-dollar expedition in the midnineties, made a final attempt to revive the dream of in situ measurement. Called project VORTEX, the mission had a grandness in scale and ambition not seen since the Thunderstorm Project, half a century before: it featured mobile atmospheric sensors and weather-ballooning laboratories, photography teams, armored aircraft taking in situ observations within thunderstorms, and a mobile probe unit on the ground. Bill Winn, a physicist at New Mexico's Langmuir Laboratory, developed the probe, known as the E-Turtle. But this, too, in all its many miles combing the plains, failed to pierce the core. Throughout the mission, just one next-generation probe managed to get close to a violent tornado. It just missed—sitting 660 yards shy of the core.

Temperature, pressure, humidity—these measurements were all off-limits to mobile radar. They were essential to lingering questions about how tornadoes form and behave, and there was only one viable way to get at them. Even so, Bluestein was done with in situ probes; and few scientists seriously considered them anymore. They'd watched Bluestein's and Winn's efforts fail and wanted no part of this work. After decades of battle, the storm had won: it would keep its secrets—for now.

CHAPTER SIX

THE COWBOY SCIENCE

Tɪᴍ Sᴀᴍᴀʀᴀs ᴍᴀᴋᴇs himself a student of the TOTO project's ill-fated history. He studies the strengths and flaws of its probe and the VORTEX mission's successor. As he conceptualizes his own device, he becomes familiar with the names, many of them the world's foremost experts, who have tried and failed to pierce the violent tornado core. He learns to look at the gargantuan odds calmly, resolutely. He is well aware that, apart from the little notice in *Commerce Business Daily*, the scientific community has largely given up on in situ probes.

But Tim is not of the scientific community. He's simply a chaser—one who has seen monster storms and their consequences firsthand. Chasers such as Tim are often all-too-well acquainted with the human toll their quarry exacts; they'll usually beat first responders to the scenes of horrific tornado disasters. Tim has witnessed the cost this past decade. He has also felt the spark of hope and purpose offered by working with Tatom's snail. Now, Tim believes he can help in his own way. He sets out to design a probe that could make history.

In his world, wind is nothing more than air rushing from high pressure to low. It's not so different from the pressure front of a blast wave. Tim's career is largely a decades-long succession of discrete blasts—

Patriot missiles, bunker busters, tons of ANFO, simulated jet-fuel-vapor explosions—each with its own distinctive waveform. Since the day he walked into Larry Brown's office, Tim has been preparing, learning to capture each gusting shock. The sensors he uses are generally the same, and they should be perfectly adaptable to a tornadic flow field.

But the real puzzle, he finds as he sets to work, won't be measuring the wind; it'll be keeping the damn probe on the ground. Despite TOTO's inelegant bulk, wind in the core can imbue even a 400-pound barrel with aerodynamic properties. If the tornado tosses Tim's hardware like a Frisbee, it won't matter whether his data-acquisition software is a work of art.

If the wing of an aircraft is the perfect form with which to harness aerodynamic force, Tim needs a shape that is its antithesis, an object that responds to an intense flow with something like the opposite of lift. The DRI team has recently been absorbed by the defense contractor Applied Research Associates, and someone at the larger shop *must* have experience with lift- and drag-resistant shapes.

Tim throws himself into probe research for much of 1998, mulling the conundrum and digging through the firm's past contracts for inspiration. The work is nothing less than the full synthesis of his skills— the exact point where the test range meets the high plains. No one with Tim's unusual background has ever shouldered the challenge of in situ probe design, much less its deployment. So the answer he finally strikes upon is one that is unlikely to have occurred to anyone but Tim.

Seven years earlier, Roy Heyman, an old-timer who consults for ARA, developed plans for an intercontinental ballistic missile launch vehicle whose shape could shed the shock wave of a nuclear device. Apart from searing-hot radiation and fallout, the immediate dangers were the blast wave's extreme drag and lift forces, which could cause the launcher to tumble and take flight. When air flows around an object, it bends and accelerates, reducing the pressure above the obstruction. If the mass of the launcher can't overcome the intensity of the flow—which in an atom bomb's shock front would be *incredibly* intense—you've got lift, just as with an airplane's wing. Meanwhile, on the leeward side of the launcher, a wake forms due to the interrupted flow, creating suction, or drag. With higher pressures at the front and lower pressures behind, the blast wave would loft, push, and flip the launcher. This is the last thing soldiers manning an ICBM need.

The trick in Heyman's design was to get the air to bend as little as possible—nothing more than a gentle deviation—which would minimize the wake and counteract the lifting force. Of all the cubes, polygonal blocks, and cylinders he tested, the humble cone proved the most obdurate. Or at least it did in theory; the project never made it past the first phase of development. For Tim, however, it presents a perfect blueprint.

If Heyman's shape can survive nuclear war, surely, with a few tweaks, it can stand up to a plains twister. Tim adopts it as the basis for a design and moves on to the next problem.

The wind's debris cloud, choked with all manner of terrestrial objects turned missile, will be equally life threatening to his probe. So, to play it safe, Tim decides on quarter-inch-thick mild steel for the shell. It has the benefit of being strong and easily machined, but it's heavy as hell. This means that, unlike TOTO, his device needs to be fairly small. The instrument must be light enough for him to carry and deploy in a hurry without herniating a disc, yet weatherproof and tough enough to survive in one of the nastiest environments on earth. This combination creates a whole new set of engineering hurdles—aggravated by the fact that Tim can't resist packing the thing to the gills with extra features.

NOAA had only requested a simple pressure recorder, but Tim has decided the task is far too uncomplicated. Like his chase vehicles, Tim's probe will amass as many gadgets as it can bear. Plus, if the device is already inside the vortex, why not collect a few other data points while there? He has in mind a complete weather station, equipped with temperature, humidity, and pressure sensors—a much more elaborate piece of equipment.

The project is an exceedingly unusual one for ARA, which typically sticks to military and national security applications over oddball weather science. But Tim is obstinate, cranking out draft after draft, each with the input of a review group of his peers. The work is far more than Tim can do alone. Cramming a twelve-bit, sixty-four-channel data-logger, fifteen pressure transducers, as well as sensors for humidity and temperature, into a space roughly the size of a shoebox is a bit like gaming out a three-dimensional puzzle. There may be only a single workable configuration.

Luckily, Tim belongs to a company with a motley pool of brainiacs. Over lunchtime bull sessions, he and the crew hammer out the details of components, overall design, time frame, and cost. To develop the miniature data-acquisition system, which will record measurements from the various on-board sensors, he taps his old buddy Bob Lynch, a software and hardware magician. Julian Lee, a young whiz from Caltech, has expertise in fluid mechanics and can make sense of the turbulent, debris-choked wind flow. Heyman, the engineer who drew up the launcher design, assists with the scale-down, from ICBM launcher to car-tire-size weather instrument.

Finally, after much deliberating and revision, Tim nails the design: a squat cone some twenty inches in diameter, and a little less than six inches tall. He calls it the Hardened In Situ Tornado Pressure Recorder, or HITPR for short. Now comes the pitch. Why should the NOAA judges award this grant to Tim, who is not a meteorologist or a scientist of any stripe? To the senior researchers behind the bid request, he will be an unknown quantity, with a wholly unconventional area of expertise compared to the atmospheric scientists with whom he's competing.

Yet that outside engineering expertise might just give Tim an edge where it counts. NOAA is looking for something that can fare better than TOTO in tornadic winds, and of this Tim and his shop of engineers have no doubt: TOTO's shape was an afterthought, while HITPR's aerodynamic shell is the product of meticulous design and calculation. On the inside, its instrumentation is research grade and built to specifications. Whereas TOTO documented conditions once

per second with the technology of its time—on paper, with a mechanical impact recorder, like a seismograph—HITPR's onboard datalogger
will sample the environment electronically, ten times each second.

HITPR is the Corvette to TOTO's Model T: a sleek update with all
the latest bells and whistles, built to shed the wind.

The real selling point, though, is what Tim believes his device can do
for science. Up to now, tornado wind speeds have been derived through
the forensic examination of structural damage. Put simply, what would
it take to bring this building down? It's a lower bound, which means
that if a record-breaking gust flattens a poorly constructed house, no
one will ever know how fast the wind really was. The surveyor can only
conclude that, say, a 130-mile-per-hour gust was more than equal to the
task. HITPR, Tim explains in his application, will provide a far more accurate wind-speed estimate by basing it on direct pressure and direction
measurements—data points that simply do not exist at ground level.

He can't overstate this point: there has always been a blind spot at
the place we most want to see. The ground level is where we live, and
in tornadoes it's where we die. Yet the tornado has remained untouchable at the surface. TOTO couldn't survive. Radar can't get there. But
HITPR can.

In the long lineage of tornado probes, Tim's is the first to be inspired by the shock wave—and the first to be shaped by an engineer
whose laboratory is the test range. Tim's gig at DRI and then ARA has
never felt exactly like work. It's more like he's been transported from
his boyhood bedroom floor and his old radios to a place where the
toys are exponentially more expensive, and the stakes are as high as
they come. But this project—his first as principal investigator—feels
different. It's something closer to a calling, as if this is what he was put
on the earth to do.

If HITPR can provide the answers to some of the enigmatic questions that linger—Is the core warm or cool? What are its approximate
ground velocities? How far does pressure fall?—then the how and
why behind tornado formation can begin to reveal themselves. Its

data could be assimilated into a tornado model along with radar and weather-balloon measurements, providing an unprecedented picture of the vortex. Perhaps one day—if the device goes into production—scientists, chasers, companies, and weather firms alike could contribute to a database. Structural engineers could have access to measurements gleaned not by educated guesses but by a finely calibrated instrument. It's a tall order to build a single-family house that can survive a Jarrell tornado, but the instrument might just give engineers a fighting chance—at safer homes, offices, hospitals. That's worth something.

As the submission deadline looms, Tim and Brown work through Thanksgiving Day of 1998 and into the wee hours, refining the proposal. Exhausted but hopeful, Tim mails his design and pitch to NOAA. Then, for nearly six months, Tim waits. If approved, HITPR will be the first project he has shepherded from conception to development. Tim is cautiously optimistic, but the guys at ARA are confident. "We thought that he would get the grant just because he put so much effort into it," Brown says.

The proposal is received by none other than Dr. Al Bedard and Howie Bluestein, along with a third senior NOAA scientist, Joe Golden. These men know better than anyone else the agony of the hunt. If they have learned anything from TOTO, it is that storm-chasing experience is absolutely paramount. If the applicant can't find tornadoes and maneuver safely around them, the experiment is doomed to fail. Secondly, at some 400 pounds and as cumbersome as an oil drum, there had never been time to deploy more than one TOTO (and in any case, only one was ever built). What attracts them to Tim's proposal is that it provides plans not for one HITPR, but for seven to ten, intended for deployment in succession. This, Bedard, Golden, and Bluestein agree, will elevate the odds that at least one might yield a direct strike. Furthermore, unlike TOTO, the turtle can easily be handled by a single man.

"There were three or four good responses, including Tim's," Bedard says. "We as reviewers thought Tim's approach had the best chance of success." Bedard and his reviewers are unanimous in their decision.

Before this process began, Bedard knew nothing of Tim Samaras. Secretly, though, he had hoped that someone like Tim would come along—someone other than a meteorologist who could chase storms and design the heck out of the thing.

In 1999, the Department of Commerce issues a $74,934 grant for Phase I development. By ARA standards, this is a pittance, but it's enough for a prototype. Tim's dream is made flesh: a fifty-pound hunk of metal and electronics that looks like a traffic cone melting onto hot asphalt. Tim paints the shell bright orange, a tribute to Tatom's snail. In homage to the most recent probe effort, where the devices were dubbed E-Turtles, Tim starts calling his device *the turtle*.

He cannot wait to test his creation, and it's from the roof of Tim's minivan that the turtle gets its first taste of the wind. Tim straps the instrument to a piece of plywood and a quad-disk pressure sensor, and floors the minivan at seventy-five miles an hour along a downhill grade near the office. The wind velocities are weak (barely F0) compared with what the turtle will one day experience—but also intensely turbulent, according to Tim and Lee's analysis. It's a good first sign.

For a stronger test, they travel with HITPR to the University of Washington's Aeronautical Laboratory in Seattle and, for a fee, place the device in the facility's wind tunnel. As smoke accompanies the artificial gale rushing over the turtle, Tim and Lee see a perfect "teardrop plume," not a chaotic wash, form in its wake. They expose it to velocities ranging from fifty-three miles per hour up to two hundred. It seems entirely possible that winds of this magnitude would send the turtle tumbling. But—as Heyman predicted—the opposite happens. Blasted with winds of over 150 miles per hour, the device holds fast. The load cells beneath the turtle even register a downward pressure. The faster the winds, the greater the downward pressure on HITPR's front edge. When Tim sees this, he knows his calculations are valid. Even better, the turtle actually seems to prefer a fluttering, turbulent flow. There is little doubt now: barring impact with a wind-driven sedan, his invention should survive a tornado.

The last challenge will be to measure the actual barometric pressure, and not some artifact introduced by the turtle's steadfast presence. With the pressure highest on the side facing into the wind and lowest at the back, Lee develops algorithms to determine not only which of HITPR's ports is recording the actual pressure, but also a solid estimate of the wind speed it indicates. This is the finishing touch before the turtle is ready to enter the wild. Then the real work can begin. "Getting an actual hit on HITPR," Lee says, will be "a whole other level of difficulty."

In early 2000, Tim travels to Washington, DC, for the second, more stringent round of funding. Before a panel of program administrators from the Department of Commerce, he defends his prototype and outlines the next phase of the turtle's development: to build a fleet of the probes, and to field them in Tornado Alley. Convinced by his pitch and the elegance of HITPR's design, the department and NOAA sign off on Phase II and cut ARA a check for nearly $300,000.

What began as absent tinkering, on the floor of a little boy's bedroom strewn with transistors, diodes, and old radios, has set Tim on a path. DRI and ARA have given him the skills. And the singular instrument called the turtle might just be the tool needed to divine the dread silhouette he first glimpsed as it churned toward Dorothy and Toto. Unlike the tumbling farmhouse, however, his probe will stand while everything else falls. If all goes according to plan, the turtles will enter a realm Tim has seen only at safe distances. This, he understands, will require the acceptance of an altogether novel kind of risk. Tatom's snail was a sawed-off 12-gauge; close was good enough. But close won't mean much to the turtle. This rifle bullet is built to pierce the heart and enter the core. Unless it does, his mission will fail.

He will have to wait until he hears the roar. He will have to watch it come on, until he can see the debris and soil lifting into the vortex. Then and only then can he activate the recorder, plant his device, and flee as fast as his V-6 will carry him. With the right approach, the right escape route, he can finally steal away with the dragon's treasure.

CHAPTER SEVEN

A TURTLE IN THE WILD

T IM HAS NEVER seemed like the kind of man to buy into provi-
dence; his engineer's brain is far too practical for fate. Yet the
following year, just as Tim is looking to test his brand-new device, opportunity again seeks him out.

The meteorological community tends to sit up and take notice any-
time someone steps forward with a credible plan to penetrate the tornado core. And in 2001, Tim's project comes to the attention of Anton
Seimon, a South African storm chaser with the backing of the National
Geographic Society's Expeditions Council. Seimon has been tapped by
NatGeo to lead a tornado-research expedition for the spring storm season. While Tim is polishing off the construction of his turtles, Seimon
is finalizing a team of leading chasers and scientists. At the helm is one
of severe weather's biggest names: Erik Rasmussen, a coordinator for
the first VORTEX research project. And leading in the field is Albert
Pietrycha, a thirty-four-year-old National Severe Storms Laboratory student researcher and expert chaser, known among his brethren by the
moniker Al-nado.

As the three players outline their expedition's goals and roster,
Tim's name finds its way onto the short list. Rasmussen is curious about

the rumors he's recently heard: "This guy Tim Samaras is doing interesting stuff, developing new probes to deploy in tornadoes." He wants to see what Tim can do.

And Seimon is familiar enough with Tim to offer a second vote in his favor. The two have crossed paths before on the Colorado plains and hit it off. They're two minds similarly obsessed, both having followed odd paths to atmospheric research. Seimon is not a degreed meteorologist either; he's a geographer by training. His specialty lies in the remote ranges of the Peruvian Andes, their ancient glaciers and unique highland ecology. But his heart resides in the bland topography of the Great Plains. Like Tim, every time storm season rolls around, he's sucked back into Tornado Alley.

When Seimon, Rasmussen, and Pietrycha reach out to Tim, he doesn't hesitate to sign on. It's as if he's been waiting for them to ask. Tim knows that HITPR, like Tatom's snail, is all theory until it enters the field of battle. After years of development, he's dying to test the turtle in a real storm. To do it with a dedicated, fully resourced expedition, led by seasoned experts, sounds like something out of a dream.

Tim's probe will be one tool among many on the Swiss Army knife of a mission. Pietrycha will be "field commander," guiding a retinue that includes Tim's tricked-out minivan, a brigade of vehicle-mounted weather stations called mesonets, a small fleet of unmanned drones, and a NatGeo film crew to capture the endeavor for a forthcoming television special. Rasmussen will guide the whole convoy from afar. With access to formidable forecasting resources at the National Center for Atmospheric Research in Boulder, he'll monitor the weather and direct the team to each day's most promising storms.

As Seimon prepares his final proposal for NatGeo, highlighting each element and composing biographies of the participants, he asks Tim for a résumé. In return, he receives a paragraph. Apart from the handwritten page he had handed to Larry Brown more than twenty years ago, Tim has never written a professional curriculum vitae.

"Where did you get your degree?" Seimon asks.

"Alameda High School," Tim replies matter-of-factly.

Seimon goes slack jawed. It has simply never occurred to him that Tim, accomplished as he is, isn't a college-educated man. Seimon briefly panics. *You're kidding me,* he thinks. *This is not going to look good. It will look like we have rank amateurs.*

But, on the other hand, who else is doing this stuff?

For its part, NatGeo loves the addition and is only too willing to underwrite the labors of a researcher whose mission is so fraught with jeopardy and photogenic drama. "It's a great story line for them," Seimon says. "The cowboy science of trying to get in front of tornadoes."

The organization approves Seimon's grant, offering funding for a monthlong effort. Tim's turtle is ultimately billed as one of the expedition's primary components. The mission's title underscores the danger the team is prepared to confront: "Inside Tornadoes: A Research Initiative." This work isn't for the faint of heart. Seimon promises "the most ambitious effort ever attempted to obtain measurements within tornadoes."

For Tim, this is an especially propitious moment. It's been three long years of grant writing, research, and development. The hunt now finally begins. From Texas up through Oklahoma, Kansas, Nebraska, all the way north to Minnesota, the team will go wherever it must to find its quarry. The unknown feels rich with promise and possibility.

It won't last.

And the experience will hold lessons that both guide and dog Tim for the rest of his life.

———

On May 20, the six-vehicle convoy embarks. They have one month from NatGeo—and a target zone of more than 100,000 square miles in which to find the swirling wind. From the very first day, every member can hear the clock ticking. They set out from Boulder toward gathering storms in far-eastern Colorado. But that same afternoon, they're thwarted, as a late cold front ices the atmosphere's volatility. They return amid drifting snow, and for the next four days, the skies are lifeless.

When the weather pattern revives, the chasing proves . . . complicated. Leading such a large contingent requires a decisive command structure. But once they reach the Texas Panhandle for their first intercept attempts, coordination issues start to plague the squadron. With cell towers few and far between, the team suffers repeated "communication lapses" at moments when they can least afford to lose contact. Having never chased together before, they have no cohesion, no common pace; the slower drivers keep splitting the team up. When they do manage to stay together, other problems sprout like mushrooms: Mother Nature isn't delivering or the team's leaders pick the wrong storms.

On May 29, they set out from Amarillo in pursuit of a broken string of cells strung along a strong Texas dry line—the classic plains setup. Chasers know to target what they call the tail-end Charlie, the southernmost cell in the line of storms, which is usually the one to cleave away, go tornadic, and thrive on an unobstructed river of fuel. That's just what Pietrycha's convoy does. They intercept the southern storm, a "stunningly beautiful, reddish-hued sky sculpture," Seimon says, and follow it all the way past the Caprock Canyons to Turkey, Texas. But it never touches down. In the meantime, they miss a mighty F3 wedge roughly a hundred miles to the north, near White Deer. A few other chasers somehow made the right call, detecting some signal in the surface weather map. Whatever that was, none of the forecasters on the team saw it.

This is the first time Tim has ever taken orders on a chase, and he finds the arrangement chafing. Every chaser makes his or her fair share of bad calls, but it's just harder to abide them when the mistake is someone else's. For an independent worker like Tim, it's like being muzzled. There are growing "tensions," Seimon says, "difficult personalities." The streak of fruitless days stretches on. As they tick by, even the NatGeo guys are fed up.

Eleven days into the mission, Tim calls a meeting in a fit of pique. With Pietrycha and Seimon gathered in his shaded Lakewood driveway, the white minivan loaded with probes and luggage, Tim unburdens himself. He says he's prepared to back out of the effort and go independent. He doesn't shout—he rarely, if ever, raises his voice—but he's forceful: The mission cannot continue this way. If he is to remain aboard, he wants greater input in the selection of targets, and the freedom to follow his own instincts. "I don't work this way. I *can't* work this way," he says. Seimon is stunned. He understands Tim's frustration as well as any, but they had all agreed at the outset to defer to Rasmussen. "He came across as a hothead," Seimon says. After a little begging and cajoling, the three strike an agreement that will keep Tim aboard and allow a greater measure of autonomy.

Following the powwow, their fortunes do not improve. Even with more independence, Tim doesn't come close to a worthwhile deployment in the first twenty days. On June 11, near Benson, Minnesota, frustrations reach a climax. At the rear of the convoy, Seimon briefly glimpses a large tornado, half-hidden in the rain. Judging by the radio traffic, it doesn't seem as though anyone else has seen it. Seimon knows he isn't supposed to occupy radio bandwidth during operations, but feels compelled to cut in. He argues that they should be chasing this confirmed tornado by sight alone. But the convoy keeps moving. Rasmussen is guiding them from afar with radar. The trouble is, the radar updates only once every five minutes or so; they're making their most time-sensitive decisions with outdated information. To Seimon, it feels as if they're driving *away* from the objective. To Tim, it runs completely counter to how he chases: if the visible facts on the ground change, respond to them.

The convoy knocks itself out of position. It's left flailing in the tornado's wake, forced to navigate around an obstacle course of fallen trees and power lines in a hopeless stab at catching a storm fleeing at forty miles per hour.

The next day, the group is in far-western South Dakota, tailing a storm that's dropping tornado after tornado—some six in all—none of which crosses a navigable road. When the storm finally passes over US 385, just south of the Black Hills, the tornado lifts, leaving them to clutch at a chaos of nontornadic winds. Even when the team manages to cohere and cooperate, it seems, the tornadoes refuse to do the same.

A week later, Tim's hopes for the season are dashed altogether. The team becomes separated, leaving Tim, Seimon, and the NatGeo crew in Nebraska for the night, while Pietrycha and the mesonets stay three hours east, in South Dakota. After 15,000 miles and a month of nearly continuous chasing, everyone is fried. The crew needs sleep, and there should be time to rendezvous in the morning.

Dawn arrives with news of potential in southern Minnesota. They may yet salvage the expedition in its final days. But as the two groups drive throughout the morning, the window shifts. The warm front that will force storm development surfaces farther northeast, in Wisconsin. If Pietrycha hurries, he'll make it just in time for the intercept. But there is no way Tim and Seimon can cover that amount of ground. To make matters worse, they've lost contact with Pietrycha again, and the guidance he's getting from a forecaster at the National Weather Service office in Dodge City, Kansas. Hopelessly out of striking range and operating in the blind, Tim ends up puttering around St. Cloud.

Only later does he learn about the half-mile wedge that raided Siren, Wisconsin. It uprooted hardwoods and sheared through homes. The town's warning system remained silent—damaged by a lightning strike just a month before—and three people were killed. The only flickering bright spot is that Pietrycha was there, in position with his mesonet team. They gathered a vanishingly rare and complete set of measurements, a scientific coup. It's one solid building block toward an eventual, potentially lifesaving understanding of the storm—a small but valuable success for Pietrycha and the expedition.

For Tim's own mission, however, the entire season is a bust. From May 20 to June 20, he has traveled some 15,000 miles and spent weeks away from his wife and children. Yet for all his restless traipsing, the turtle remains untested. Tim limps back to Colorado and returns HITPR to its place in the basement.

In the off-season, a notion begins to harden into conviction in Tim: he is a chaser who doesn't need another man's forecast to find tornadoes. He vows to trust in his own cunning beneath the storm. He won't tether himself to a cumbersome entourage, no matter how decorated. The season is too short and too precious to hand away. One chaser, with his foot on the gas, is worth twenty experts arguing over directions.

CHAPTER EIGHT

THE TOREADOR

B UILDING A DEVICE that will remain immobile in winds strong enough to hurl railcars was the easy part. For all the years Tim has handled weapons designed to destroy on massive scales, the testing has always been meticulously controlled, each variable understood and accounted for. High explosives often require a deliberate dose of energy to detonate. There are fail-safes in place, strict layers of firing-system checks. Warheads may contain awesome power, but that power is unleashed only on his mark.

There are good reasons why the quest Tim is now undertaking has never been realized. It's not just the messiness of a large-scale chase. Even when Tim has been able to chart his own course, he still operates in a world that is, by its very nature, in flux. Unlike on the test range, he controls nothing beneath the storm. Every object is a missile, every passing telephone pole a potential crushing blow, every dirt road a quagmire. There's a thin line between too close to the tornado, and too far. His life—and the fate of his mission—depends on knowing the difference, and on minimizing every infinitesimal variable within his power.

On February 16, 2002, Tim meets with Seimon at an IHOP in Lakewood to discuss a new NatGeo grant proposal. The pair now imagine a stripped-down mission, its scope winnowed to a single objective. No mesonets, no instructions phoned in from Boulder, no unmanned drones—just Tim's probe and perhaps a small media crew from National Geographic. Tim argues that the mission's command structure should likewise be simplified. Strong-willed chasers are bound to argue to the point of paralysis over which target to pursue. One person, Tim says, should be charged with making the ultimate decision when there is no consensus. And Tim believes he should be the one to make the call.

As they settle in at the table, he hands Seimon a single page of typed notes—a manifesto of sorts—that he proposes should govern field operations going forward. "This mission is ALL ABOUT getting In-situ measurements with no allowance for other programs to interfere," he writes. In other words, they want to make history: pierce the heart of the tornado or bust. He continues: Each morning the team will convene to discuss the forecast, but "being that I am responsible for the fielding of these probes, I have the final decision if there are any differences of opinion." The manifesto strikes a surprisingly, uncharacteristically authoritarian tone. Yet Tim must have the rigid hierarchy of the previous mission in mind when he writes in one final caveat, "There will be no 'dictating' on the target. The subject is always open to suggestion and review."

Since last year, Tim's attitude has hardened. He's a more focused, more intense man. He's been hard at work, building eight turtles, which he plans to deploy at intervals to sample multiple transects through the tornado. Seimon gamely offers to take several, reasoning that a second deployment team will increase the likelihood of a hit. But Tim flatly refuses to part with even one. His chasing long ago crossed from hobby to obsession; now it's almost as though his fate and that of the turtle are bound together. The animating cause of his life is the fervent belief that his invention will be the exception, the breakthrough. If anyone is going to make history with the turtle, it isn't going to be Seimon. It can only be Tim.

Through February and into March, Seimon corresponds with other scientists who might round out the mission's forecasting capabilities; though he considers himself a specialist in High Plains storms, he's still occasionally baffled by those to the east. In late April, however, Seimon gets bad news: the Expeditions Council of the National Geographic Society has declined to fund their grant proposal.

It's a blow, especially so close to the start of the season, especially since Seimon's work still feels unfinished after failing to "land the big fish" last year. Still, he vows to help Tim accomplish his objective in whatever way he can. The two have developed a strong rapport, and Seimon admires Tim's commitment. After regrouping, they decide to string together a series of chases out of their own pockets. All they really need is a driver and a navigator, and a little money for gas and motels. They wanted simplicity, after all. Now they're going back to basics.

For Tim, the loss of NatGeo contains something of a silver lining. There's no hierarchy, no holder of purse strings. He has always cut an odd trail of his own invention. This doesn't seem so different. The turtle and the mission are in his own hands: he is free to charge at the core as he sees fit.

Through the spring and early summer of 2002, Tim and Seimon fall into the grinding rhythms of storm chasing—a week on the road here, a couple of weeks there, whenever and wherever the weather models betray a glimmer of hope. Tim has saved up his vacation time at ARA, and if he needs an extra day or two, he pulls weekend shifts to stay within the bounds of company policy. On some days, Seimon rides along. When he can't come, Tim invites his friends—guys who can spare a few days to hold a camera, split the cost of a motel room, keep him company, and raise his spirits when things look bleak. To a one, they believe in his mission and have been infected with his passion. The roster includes colleague Julian Lee, Tim's neighbor Brad Carter, and his brother-in-law, Pat Porter.

It isn't quite like Tim and Porter's previous carefree pursuits in Lipscomb County, Texas, and the like. Tim is chasing harder now. He keeps to the road with religious zeal, staying out longer than he ever has. One by one, his companions cycle through, unable to keep pace. They have to return to their jobs and families. Tim is the only constant. ARA gives him as much space as it can; the crew knows how badly Tim wants to prove HITPR. Within the Samaras household, too, Kathy recognizes that a new drive has taken ahold of Tim. "I could see how excited he would be about it, and how interested he was," she says. "He felt like he was doing good." The lengthy work trips for DRI and ARA had long become a part of the household rhythm. His life on the road, Kathy says, "was part of who he was." Now, that part of him is engulfed in a singular hunt for the storm.

As the months wear on, though, the distances feel longer and wearier. Due to the paucity of meal options in sparsely populated regions where tornadoes most often occur, Tim is forced to subsist on grub at greasy diners, or to scavenge food from the nearest Allsup's convenience store when nothing else is open. He then calls Kathy and beds down for the night in some down-in-the-mouth motel with holes in the drywall and a dead cockroach awaiting him on the bathroom floor. If the next day's target is far away, he catches only a few hours of fitful sleep before it's time to hit the road again.

Over unspeakable miles of flyover country in a single season, the long sedentary hours exact their toll. Some chasers joke that they can practically feel the clots forming in their legs. Tim's companions marvel at how ably he weathers the chase (and maintains his trim physique). The missed tornadoes and busts are the only vicissitudes that sap his resilience.

To Julian Lee, Tim reveals an entire subculture Lee scarcely knew existed. At some depopulated crossroads, they'll find a dozen or so chasers who have all reached the same conclusion about the weather to arrive at precisely the same place. They pass the time as most chasers do, speculating on the timetable for storm initiation, studying road maps, plotting target locations, and telling war stories about storms of the past. Before long, everyone is facing the sky, appraising cumulus towers as though judging prized cattle. The "crispier" they look, the better. Inevitably, a debate breaks out over the strength of the tornado that has yet to form. It could be augured by a particularly curvaceous hodograph—the line trace that represents wind shear. Or it could all hinge on the penetrability of the cap, the layer of stable air that acts like a lid on storms.

Eventually, Mother Nature puts a stop to the arguing. The cap will break. The cumulus towers will become cumulonimbus mountains. And the chasers will light out, leaving contrails of dust down back roads.

Until Lee began chasing with Tim, he never fully understood that he already spoke the mathematical language of storms. Fluid mechanics is his specialty, and in the sky it finds one of its grandest expressions. The explosive, hundred-mile-per-hour vertical growth of towering clouds is an awe-inspiring manifestation of the simple tendency for warm, moist air to rise. The sky, he now sees, is an ocean of latent energy. Life depends on the benign expenditure of that energy. Yet a rare process can transmute it into a knife's edge capable of horrific carnage. Chasing "gave me respect for how much energy is in the air we breathe," Lee says. "It's something I'd never worked out until talking to Tim and those guys. There were a lot of things I'd learned in textbooks about turbulent flow, wakes, and vortices that I could see in the lab, but I didn't realize they happen at such a large scale in nature." Out here, the experiments span state lines, evolving from one moment to the next with infinite variation.

The other thing Lee learns on the road is just how difficult it is to find tornadoes.

Tim misses an outbreak stretching from the Midwest to the Mid-Atlantic in April 2002. Six people die, and dozens of tornadoes, including a powerful F4 in Charles County, Maryland, cause hundreds of millions of dollars' worth of damage. But the hills and forests of the Midwest and Mid-Atlantic are beyond his territory; Tim requires flat land and predictable, gridded roads on which to navigate.

On May 7 he gets a near miss on a probe—closer than he's ever been—outside Pratt, Kansas. The nearest turtle measures a twenty-four-millibar barometric-pressure drop and a peak wind gust of seventy miles per hour at the northern edge of the funnel—it's so close he could have hurled a rock into the tornado. But close isn't enough. His probe must pierce the heart. "There's a hell of a lot of disappointment," Carter says. "You spend all that time and effort, and then you miss it, just by a little bit."

But failure, in this case, isn't without value. Tim is learning that to accomplish what no researcher has before, he must get closer than any has dared. Like the toreador who waits until the last moment to pivot from the bull's horns, Tim will have to stand in the path just before escape becomes impossible. Carter isn't sure whether this is a good thing. But whatever Tim's reservations, he has now grown comfortable in proximities that would have terrified him a few years ago.

For all the many miles Tim and his ragtag crew cover, again, the 2002 season ends without success. Tim retreats to lick his wounds and gear up for next year's battle. But the storm will not wait; fall holds a surprise event in bad terrain. In the early afternoon of November 9—an unseasonable month for an outbreak—the first tornado touches down in Arkansas. And they don't stop coming. Dozens rake across the South and up through the lower Ohio Valley. One out of every six is what NOAA refers to as a "killer."

At 5:40 p.m. on November 10, there are simultaneous outbreaks occurring in an almost-continuous line of supercells from Louisiana to Lake Erie. The mile-wide wedge that enters Van Wert, Ohio, tosses a car through the wall of a movie theater. It lands on empty seats in the front row. Another vortex comes to Mossy Grove, Tennessee, in the night. The people who live there had believed Lone Mountain would shield them. But street after street is mounded with rubble, and five of their own are dead. In the town of Clark, Pennsylvania, the body of an eighty-one-year-old man is discovered in his basement, buried beneath the ruin of his home. Clark had received only a severe-thunderstorm warning, six minutes before the tornado arrived. The tornado warning came two minutes after the twister was gone.

By the time the sun rises on the eleventh, Veterans Day, thirty-six have been killed. Deadly, destructive, unpredictable. Tim has been in the presence of the fastest wind on earth more than almost anyone else alive. But when a tornado leaves the empty fields and enters a town, he is still astounded by the way the air—without an igniting charge or an explosive compound—can act like a bomb's shock front. Sometimes people know it's coming, and sometimes they don't. In either scenario, the best we can do is hunker down as people always have and hope the wind will miss. A chaser can either stand by, or he can do something about it. Tim doubles down on the probe.

CHAPTER NINE

STRATFORD, TEXAS

THE 2003 TORNADO season arrives like a reset. A new year, fresh with possibility. Tim is able to venture back out instead of watching helplessly from afar. He keeps a close eye on the predictive weather models, as they show early promise. Then May 15 dawns like a gift from the storm gods.

The Texas Panhandle sky is flooded with combustible atmospheric fuel from the Gulf of Mexico, and the unstable mass is on a collision course with dry western air and a howling, fifty-knot current moving east at 18,000 feet. Tim knows only that something could happen, not that it will. But the potential is enough to prompt him and Anton Seimon to pile into the minivan and depart from his Lakewood driveway at nine in the morning.

The road trip to the target is a rump-numbing haul no matter what. It's the sound track, though, that makes it feel like an eternity to Seimon. Gone are the Clapton CDs that were once in heavy rotation. Tim insists they listen exclusively to the Weather Channel, which he streams via satellite. Through Colorado and into Oklahoma, the saw-toothed mountains melt into the southern plains to Muzak, "this Kenny G stuff," Seimon says, punctuated by on-the-hour weather updates.

In the heat of the chase, Tim communicates on only one frequency, and this single-minded enthusiasm either infects a man or exhausts him. Tim inveighs endlessly about the mission, his customized chase vehicle, and epic storms of yesteryear, as though there were no life outside the chase. By his way of thinking, they're currently bearing down on a storm that's about to expend a nuclear warhead's worth of energy. What else could possibly be as interesting?

In the many thousands of miles that Tim and Seimon have traveled together over the last two years, the pair have spoken surprisingly little about family or work—delving neither into Tim's explosives expertise or Seimon's expeditions into the Andes. It's not that Tim is uninterested or intensely reserved. On and off the road, he's one of the most genial guys any of his friends know. He's generous with advice for the newbies who look up to him, and a cheerful troubleshooter for any chaser whose ham radio is on the fritz. But underneath all that there's an undeniable edge to Tim—and it rises to the surface just when cumulus clouds harden into anvils.

In recent years, skepticism has been growing about the viability of Tim's mission. Folks such as Erik Rasmussen had demonstrated initial excitement, and even offered assistance. But as more seasons pass and "close" is all Tim has to show for his efforts, the meteorological community's curiosity fades back into its prior cynicism. Tim has something to prove. To the doubtful academics and researchers, to anyone who questions whether a mere chaser can bring home storm science's holy grail, Tim wants to show off just what he can do.

Seimon understands the frustration—"We both have had to swim against the tide at different points," he says—and wants to help Tim channel it. He's ready to stick with Tim, even through the Muzak. Both men are "absolutely mesmerized by the atmosphere," Seimon says. "And when you're clear about that, everything else is a detail to be worked out."

With the windows down, the pair enters the Texas Panhandle under a rapidly graying afternoon sky. Tim and Seimon listen to the moaning of power lines in the wind, savoring the subtle shifts in pitch as the gale slackens and swells. The stage is just about set. A line of thunderstorms now develops before them, stretching south to north, from Dalhart, Texas, to Boise City, Oklahoma. Within a few hours, fifty miles of super-cells will pop up along the dry line, strung like pearls.

———

At a little before six that evening, Tim's chase vehicle enters the cross-roads hamlet of Stratford, Texas, one of those lonely, wind-scoured out-posts afloat on an unbroken ocean of grain. Bewhiskered by quivering antennas, and the white orb of a portable satellite dome, the Dodge Caravan cruises down an empty US Route 287.

Clouds as dull as slag have choked off the light of a late-spring sun, bringing an early dusk to the Panhandle. The asphalt is wet and shining like obsidian from the passing storm. Tim guides the minivan through town, craning his neck for a glimpse of the western horizon through the gaps between trees and the little houses with dusty hardpan lawns.

That's when he catches sight of it: a lowering too deep to be the wall cloud of the mesocyclone, too solid, too big, too well-defined, for a false-alarm "scudnado." Though Tim glimpses the shape only for an instant, his every instinct signals tornado. Seimon doesn't see it yet, off to the northwest, but he knows Tim isn't in the habit of crying wolf. He begins filming through the windshield for the scientific record. "Time: 22:45:21," Seimon narrates. "We're going to take a look."

Stratford's residential sector sweeps past, its water tower, then a se-ries of corrugated-steel buildings. Tim accelerates, and the minivan's motor thrums. At a break in Stratford's meager skyline, Tim spots the silhouette at last. "Oh, my God," he says. "We've got a wedge on the ground."

"Wedge on the ground," Seimon repeats, though he can't quite see it through the rain-bleared windshield. "Wipers, please." The blades slash across the glass. He pans the camera frame over the northwestern horizon, where the low cloud bank hovers parallel to the smooth plane of the prairie, then suddenly plummets to earth. There it is, the telltale funnel. "Oh, Jesus," he gasps. "Wedge tornado on the ground."

They race northwest out of town down 287, and ten minutes later Seimon is filming through the driver's window. Beyond Tim's nose, he sees an inky mass of shifting shapes tickling the endless drip-irrigated cotton tracts. A series of vortices ride the outer rim of a broad tornadic circulation before fading, replaced by other thin twisters. This is a classic multiple-vortex tornado. At a certain point, as its strength oscillates, there seems to be nothing out there but a lazily swirling fog, a ghostly carousel. Then it builds and darkens again.

Soon, they are facing the darkness directly, within minutes of its most powerful winds. As they enter the rain curtain at the front of the storm, the tornado drifts neither left nor right. Instead, it only advances, growing ever larger. The beast is coming for them. Or, rather, they're coming for it.

Tim and Seimon now leave behind even the most audacious chasers, parked along the highway, some hunched over tripod-mounted cameras. Tim advises Seimon to prepare himself for the hail core, which runs up against the outer edge of the updraft. They can feel themselves cross over the instinctive margin of safety and into dangerous territory. They are entering no-man's-land, the place considered too close, a violation of storm chasing's cardinal rule. Once, near towns like Last Chance, Colorado, Tim obeyed the one rule. He kept his distance.

Now he is placing himself in the crosshairs of a tornado, and what strikes him and Seimon most is that this is no accident. They haven't lost sight of the thing in the rain or misjudged the course and strayed into its approach. In fact, they have predicted the tornado's path with rare precision. It's not so much frightening to find themselves so close as it is surreal. What they're doing isn't storm chasing anymore. It's something else. Tim knows they don't belong here, but this is what it takes.

The snap of baseball-size hail against the roof brings them back to the reality of their position. "Oh, my God, that was, ah . . . Tim, you've gotta get out of the car in this. Be careful."

Tim sees the tornado churning toward their stretch of highway, closer now and maintaining the same trajectory. Apart from the hailstones large enough to kill, the duo are in perfect position for a turtle deployment.

"Ready?" Tim says.

He swings the minivan around into the opposite lane, the cabin resounding with the erratic tattoo of heavy ice against metal. "Watch your head, my friend," Seimon cautions, and looks out onto the approaching circulation, a diffusion of drifting cloud and tightly spun vortices. "Tim, it's very close in." The vortex is approaching at nearly thirty-five miles per hour. "You're in position. You're in position."

Before Tim can open the door, there comes a startling thwack against the roof, heavier than before. "Oh, shit," Tim cries. "That was huge!"

"I don't know . . . shit," Seimon says. "You can't go out."

Tim ignores him. "Let's go."

"Okay . . . debris half a mile and closing."

Tim drives roughly fifty yards back to the south in an attempt to escape the hail core, watching for the left or right drift. He slams the brakes. "Okay. Let's go."

Seimon begins to step out, but the roof of the minivan rings with another deafening impact. "Oh, shit!" he cries.

"God damn," Tim shouts. *"Woohoo!"*

He crawls from the front seat into the back, lifts a floor panel, and begins to extricate a turtle. Tim grimaces as he hefts the tire-size object. "How are we doing?" he asks

"We've got *baseballs* falling." Seimon winces at the pearlescent orbs streaking to earth. "These will take your head off, man. Just sling the thing out! Winds increasing slowly. Watch your head!"

Tim ducks out of the van, holding the turtle high above his shaggy crown of hair as a shield. Hailstones thud dully against the grass all around him. He drops the device some fifteen feet from the minivan. He casts a quick glance at the approaching tornado then lifts his hands defensively as he runs back to the vehicle. *"Watch your head, watch your head, watch your head!"* Seimon shouts.

The camera registers the sound of the sliding door slamming shut.

"Can you hear it?" Tim says, unmistakable glee in his voice as he lunges into the driver's seat. He's not talking about the thud of hail now.

A low-amplitude roar emanates from the tattered clouds spiraling toward their location—it's a steady, unstoppable crescendo. "Holy, Jesus, can I ever *hear* it!" Seimon says. "Okay. Time: 22:58:17. GPS set point."

The minivan tears along the highway as the rain overtakes them, and the wind screams out of the south. Seimon looks over at Tim. He is clutching the steering wheel as though his life depends on his grip. They can barely see the road ahead.

"You might want to slow down a little bit," Seimon urges. But Tim doesn't let up, even as he struggles to maintain control of the vehicle.

"No." His voice betrays the shadow of an emotion that doesn't often cross Tim's confident exterior.

The wind intensifies, buffeting the minivan over the lanes, driving into them broadside like a lowered shoulder. The right side of the vehicle is beginning to lift perceptibly.

Tim believes the tornado is about to overtake them.

"We're gonna die," he says, gripped by a pure animal fear.

He sounds like a man who has just discovered his own terrible mistake. They have pushed too far, tempted fate. All that's left now is to learn the cost of violating chasing's one rule.

Seimon tries to reassure him: What he's experiencing isn't the tornado itself. It's likely the outer circulation—too close, but still escapable.

The minivan tunnels blindly through walls of violent gale-driven rain. The bright-yellow highway dividing lines fade beneath gray currents. "Oh, my God," Tim mutters. He shoves the accelerator to the floor.

"Slow down, Tim," Seimon urges. "Slow down. Slow down. Winds at one hundred miles per hour."

To the right, a power pole cants into their lane. More fall. Others bow like pliant saplings, looking as though they might topple into the road at any moment. "Power lines down. Left side," Seimon directs. "You don't want the power poles to come down on you."

Tim swings into the oncoming lane to avoid them.

"Beautiful," Seimon cries. "Winds easily over one hundred miles per hour now. The power poles are bending! I've never seen that in my life! They're *bending*. More power lines down. Slow down. Slow down."

The headlights of cars shine in the gloom ahead, behind a series of looming poles and their arcing transformers. Tumbleweeds skitter before them. Then, as quickly as the rain curtains had enveloped them, the sky clears. The light pales again. Rain falls gently.

Tim and Seimon regard each other with wide eyes and manic, adrenal grins.

"You did it," Seimon says. "Well done, my friend. Well done."

"That was fucking close," Tim sputters.

"That was beautiful," Seimon says, as Tim finally begins to throttle down, to relax his grip on the wheel. "That was perfect."

———

As soon as the storm passes, they navigate back down the highway, now strung with electrical lines, past fields littered with power poles. Tim gets out and strides over to his turtle. He fixes the conical point between his knees and pries open the lid covering the data recorder's switch. The red light strobes. He looks to either side, at the nubs where telephone poles have been snapped a foot or two off the ground. "We've got telephone poles down to our north, and lots of telephone poles down to our south," he observes.

His heart must still be hammering in his chest, though he sounds calmer now. They crossed a line today, and he knows it. The tornado was weakening when it reached them, but at any moment that could have changed. The 100-mile-per-hour current they struggled through might have intensified into a 150-mile-per-hour gust strong enough to batter the minivan into the ditch and an end-over-end roll.

Yet as Tim balances the turtle between his thighs and deactivates the data recorder, all seems forgotten. The fact of their escape eclipses the terror of the moment. They made it out alive, with one hell of a story to tell. On to the work at hand.

Whether the tornado core passed over the turtle will become a point of debate. Based on video provided by other chasers, Tim maintains that it did. Others, most notably Joshua Wurman, the founder of the Center for Severe Weather Research, say that HITPR caught an oblique slice of a dissipating-though-dangerous tornado. Nonetheless, Wurman, one of the most prominent atmospheric scientists in the field, wants to include Tim's data in his mobile-radar analysis of the same storm. For the first time, the turtle is yielding information that's useful to other researchers.

In the paper that Tim coauthors with Wurman, he describes a forty-five-millibar barometric-pressure drop, indicating the passage of intense winds—as well as a series of peaks and valleys in the pressure trace, consistent with the movement of several suction vortices. The measurements are encouraging but by no means groundbreaking. Bill Winn, a VORTEX researcher and physicist at New Mexico's Langmuir Laboratory, had gotten a probe within less than half a mile of a violent F4's center in Texas eight years before.

Nevertheless, Tim is getting closer. Twice now, here and in Pratt, Kansas, he has succeeded in getting his turtles in front of tornadoes. They just haven't been hit head-on. Yet.

That's precisely what starts to worry Seimon.

After the close call in Stratford, he decides that he will not venture back into the path with Tim. Reluctantly, he withdraws from the mission. He has grown rather fond of Tim over the years, and he's still inspired by the dream of a historic intercept—but Seimon can't justify the risk of entering "no-man's-land" again. After several years of working near tornadoes, they have no horror story to tell. They're getting good at this, and they came out of Stratford without so much as a scratch. But it quite easily might have ended differently. The tornado ran over that turtle a mere eighty seconds after it was deployed.

It's simple statistics, Seimon believes. Whether their chances of getting hit on a given deployment attempt are twelve percent or even just two, if they play the odds long enough, eventually they will lose. "You can only roll the dice so many times," Seimon says, "before things go wrong."

He turns to other plans instead. He'll go back once more to the unexplored glaciers of the Cordilleras. He'll finalize the research for his long-neglected thesis. He has met a woman named Tracie, whom he plans to marry. He doesn't intend to die before the wedding.

Tim doesn't tell Kathy about how close he gets in Stratford, or how close he needs to be to pull this quest off. He may have stumbled too close, he'll admit to himself; but he still thinks he can find the right balance. Each storm is teaching him something. They might not be yielding the right data quite yet, but they're still showing Tim their tricks. A chaser—like a lion tamer—learns only in dangerous proximity. He'd never be able to voice the lessons, but he's developing an animal sense of how the twister moves, how it evolves, when he must flee, and when he can pounce.

CHAPTER TEN

MANCHESTER,
SOUTH DAKOTA

I N EARLY 2003, Dr. William Gallus of Iowa State finds himself facing an intractable problem: he needs something that doesn't exist. In 2001, the professor had broken ground on a new tornado simulator. The machine was to be one of a kind, powered by an eight-foot fan mounted inside a cylinder the size of a subway tunnel. The entire apparatus would be suspended from a five-ton crane inside the university's Wind Simulation and Testing Laboratory. Not only was it designed to generate wind speeds of up to sixty miles per hour, its vortex would eventually be capable of lateral movement.

An academic with a smooth, boyish face, Gallus is tantalized by the structural-engineering insights the simulator might yield as it translates over model buildings. He envisions revelations that will lead to a fuller grasp of how tornadic winds interact with structures, and how that might ultimately help us to build houses that are more tornado resistant.

The construction project itself, made possible by a National Science Foundation grant, is now complete. But before he can learn anything from it, he has to be sure that the simulation resembles the real thing.

The finest data set he can locate belongs to Joshua Wurman, drawing from his work with mobile Doppler. Wurman succeeded in scanning the entire life cycle of a tornado near Spencer, South Dakota, in 1998, and his data amounts to a high-resolution X-ray of its internal structure and evolution. Only a few years ago, nothing of the sort would have been available to Gallus; Wurman didn't build his Doppler on Wheels (DOW) until 1995. Now, Gallus is able to compare the Spencer storm to the wind-speed distribution in his simulator, and the results are favorable.

But there is, for the professor's purposes, one notable limitation to the DOW. Radar operates on line of sight. With distance, the beam becomes obstructed by everything from buildings to the spherical curvature of the planet. As a result, it can only collect data at tens or even hundreds of meters above the ground. Gallus still lacks any way to validate the wind profile at the simulator's lowest levels—where people live, where houses stand, where he plans to place his model buildings. He's come up against the same problem that has stymied researchers for decades.

He knows there is only one way to gather such data, and he certainly has no intention of attempting to do so himself. He's well aware that the danger and difficulty inherent in probe intercepts has been a stumbling block from TOTO to VORTEX. In most quarters of the scientific community, a probe intercept has come to be considered quixotic, likely reckless, and almost certainly impossible.

The two aerospace engineers who report to Gallus have only a limited awareness of this fact. What they do know is that the profile of the vortex is incomplete. Without measurements from the boundary layer of an actual tornado for comparison, their simulator is based on theory—a glorified guesstimate.

"You're asking for too much," Gallus says when they push him. He doesn't know what the solution is, but he knows a dead end when he sees one. "We don't have wind information at the ground," he says. And they probably never will.

That's why Gallus can't help but chuckle when he reads an email that has been forwarded to him by the Iowa-chapter president of the

National Weather Association. Gallus and his colleagues have been busily rounding up guest speakers for the Seventh Annual Severe Storms and Doppler Radar Conference, scheduled for March 2003 in Des Moines. A man named Tim Samaras has written the organizers and made an offer that, to Gallus, betrays more than a little hubris.

Tim has essentially invited himself to the conference, and he's asking for the chapter to cover his costs. If so, Tim proposes, he will present data he has collected from tornadoes using his turtle probes. He confesses that, as of this writing, he has yet to place one directly inside the core, but Tim is confident he will do so soon.

The email gets passed around among the leadership at NWA. That a prospective speaker has invited himself is the least outlandish aspect of the proposal. It is the nature of his mission that raises eyebrows. Among seasoned field scientists, stating one's intent to deploy a probe inside a violent twister will surely be met with reflexive skepticism. Especially when the proposer is unassociated with any of the top universities or private research organizations. Gallus and many others have never heard of this Samaras fellow. Yet, standing in the shadow of the field's top minds, he proclaims that *he* will accomplish what they never could?

Tim has been told before that he is wasting his time, that the probe intercept is impossible. But even though they have their doubts, Gallus and the others can't deny the brief pang of excitement at this sign that the dream has not died off completely. Because if by some wild chance Tim does manage to prevail, his data will be priceless.

As Gallus finishes reading Tim's email, he can't help thinking, *This guy is a yahoo.* Yet Gallus has to admire his pluck.

———

Months later, in a cheap motel room, Tim gazes into the screen of his laptop with road-weary eyes. He and Pat Porter are up again at the crack of dawn, surrounded by a metastasizing accumulation of dirty laundry, in another tiny town somewhere in Nebraska. Tim scans an alphabet soup of acronymed weather models, searching for some clue above the

Great Plains. The men scarf down their breakfast—a diner staple of eggs, biscuits, and bacon—as the day's atmospheric variables flip over in Tim's mind like a Rubik's Cube. He's anxious to get on the road.

Storm chasing is a gamble, and for several weeks now Tim has played the odds, wagering thousands of dollars in gas, lodging, food, and an obscene number of miles ticking ever higher on the odometer of the family Dodge Caravan. He has spent his days behind the wheel, watching the hours transform the land. In theory, the objective is simple enough: find a tornado, get in front of it, drop the probes, and move the hell out of the way. In practice, Tim has come to terms with the fact that he is hunting ghosts. He got close in Pratt. He got close in Stratford. But the last few years have been a litany of busts and near misses. Perhaps he was wrong to believe that anything sets him apart from past hunters. Maybe the biggest issue with TOTO wasn't the probe but the foe—unpredictable, untouchable, unbeatable.

Another season is drawing to a close. The date is June 24, 2003, and a storm-killing high-pressure ridge will soon smother the plains with hot and bone-dry weather. The tornadic activity is shifting north, into Canada.

Added to nature's own encroaching deadline are the financial considerations. This year, he was able to get the National Geographic Society to underwrite his campaign to field the turtles, but without strong results there likely won't be a second grant. This is the last day *National Geographic Magazine*'s embedded photographer, Carsten Peter, can remain on the road with Tim. Peter has already begged his editor for two extensions; the clock has run out.

The small team finalizes the day's target, and at around eleven that morning, they step outside into a warm, gusting wind out of the south. Slate-bellied clouds give the sunlight straining through a dingy cast. Tim and Porter climb into the minivan and head north; the photographer and his guides, Gene and Karen Rhoden, a husband-and-wife storm-chasing team, follow behind in an SUV. They plot a course for the Nebraska–South Dakota border, picking their way along a tangle of state roads and federal highways through the gently undulating, grassy dunes of the Sandhills. That afternoon, they are treated to their own private air show: a pair of dogfighting jets from Offutt Air Force Base dive and bank and launch flares that streak across an acetylene-blue sky.

Tim doesn't say much on the drive. As he steers for South Dakota, he begins to second-guess himself. Rhoden and the others had pressed for a play on the northern system of storms, which are likely to form somewhere near the border. Tim agreed, but he knows it risks missing an epic tornado farther south, in central Nebraska, Kansas, or Oklahoma. He's been burned this way before. Normally he wouldn't beat himself up—or even give doubt room to creep in. But Carsten Peter has just canceled yet another flight home to stay on, and NatGeo has invested thousands of dollars in his mission. They're expecting . . . *something*. Tim knows his dream intercept is possible. Getting it is simply a matter of finding the right storm. On this day, though, the sheer size of the plains seems especially daunting.

A few miles after crossing the Missouri River into South Dakota, they gas up and prepare for the chase. The Storm Prediction Center in Norman, Oklahoma, has issued a tornado watch. Portions of Iowa, Minnesota, Nebraska, and South Dakota are under the gun. Particularly around southeastern South Dakota, the weather service describes the atmospheric instability as on the "extreme" end of the scale. It looks like the Rhodens were right—and, moreover, like there's serious potential today. Once the sun begins to sink toward the horizon, an eighty-mile-per-hour jet stream will kick in from the west, and the storms will quickly intensify from severe to tornadic.

The harbingers are all around them as they head north: cumulus towers form, white as bleached cotton and smooth as polished marble; they loom like precipitous atolls over a cerulean sea. This means that warm, moisture-suffused air—TNT as far as the chasers are concerned—is convecting toward the upper-level winds that will transform gentle giants into glowering supercells.

By six that evening, thunderheads congeal into a line stretching

from western Minnesota across the southeastern corner of South Dakota, and into Nebraska. As projected, they grow explosively in this
charged environment, erupting into the lower atmosphere like a cal-
dera's rising ash column. Within twenty minutes, the first tornado
touches down.

At 6:16, Tim starts tracking a northeast-bound twister near Woonsocket, South Dakota, a former rail junction in the melon and wheat
country east of the James River. They'll need to drive fast to get ahead
of it, but this is a storm that could redeem the season. Roughly a mile
out, they behold the prototypical vortex: its funnel is gracefully tapered, its hue ever changing with the angle of the light. When Porter
zooms in on the tornado at ground level, the camera reveals staggering
violence. Scarcely detectable suction vortices lick out of the earth like
tongues of fire and vanish almost as soon as the eye can register what
it sees. With the sun behind the plume, the particulate looks as black
as coal dust. Judging by the way this tornado chews through windbreak
trees like a wood chipper, it is deadly, by its nature unpredictable—an
exceedingly nasty specimen for the turtles if they can get ahead of it.

Tim approaches from the east, then swings north onto a dirt road,
racing parallel on a path he hopes will eventually bring them into intersection.

"Okay," he says, "let's get ready."

He guns the minivan over the narrow South Dakotan back road,
which is wide enough to admit only one vehicle comfortably. His eyes
dart between the terrain ahead and the slender column scoriating the
crops. In the camera frame is Tim's recognizable hawk-nosed profile.
The tornado is just off to their northwest, but Tim is gaining on it. In
another moment they're dead even. "I gotta wait until I get the right
angle on it," he says. The tornado continues to move steadily to the
northeast. He just needs to guess where it will cross.

"Help me with some roads," Tim instructs. Most likely, he wants to
be certain that the road doesn't dead-end ahead, as these farm lanes
occasionally do.

Porter consults the computer monitor displaying the DeLorme road map. "You've got an east-west coming up right up here," he says.

"I wanna go north."

"You're fine north."

Within half a mile, Tim begins to brake. He glances over his shoulder toward the tornado to his eight o'clock and brings the minivan to a halt. He steps out onto a rust-red gravel road, wearing jean shorts hemmed at the knees, white socks pulled to his calves, and a sweat-stained Henley T-shirt. Porter rounds the front and resumes filming. The sound of it never fails to bring him up short, as though all the winds of the world are converging on a single point, here in South Dakota.

The twister has kept to the fields the entire chase, and this may well be the only road it crosses in its life cycle. But now, with sight lines unblinkered by the minivan, Tim can detect in its behavior all the signs of a tornado in terminal "rope-out." This is the phase in its development during which the funnel contracts, and the trunk begins to wander like a crooked vine until its imminent death. The trouble with deploying on a vortex in this end stage is that these are its most erratic moments. There is often a certain amount of stability found in a tornado's maturity, a kind of straight-ahead churning. A roping tornado, however, is like wildfire—a small change in the wind and it may veer unpredictably. Still, if the twister doesn't dissipate first, there's a chance it could pass over Tim's current position with plenty of room for escape. If this is the day's last gasp—maybe even the final tornado of the season—he's sure as hell going to deploy on it.

Tim pulls a turtle from the minivan and hesitates near the tall grass at the edge of the dirt road, the conical shell braced against his abdomen. He's watching, waiting for the tornado to make the next move. He props the probe on its rim, flips the activating switch on the underside, and carefully lowers the device onto the gravel.

He dives into the minivan to retrieve a second probe, stowed upside down, its point secured by a hole cut into the floorboards—this he places some twenty yards down the road. "All right, let's go," he shouts, sprinting toward the minivan. "Let's *go!*"

But as soon as they begin to drive away, the funnel fades. Porter can only see the disembodied tantrum of soil and grass whipping at the surface. "Ah, I think it's dissipating," he says.

Tim won't believe it. "Not yet!"

Another hundred yards down the road, Porter can scarcely detect the surface-level rotation.

Tim brakes hard. "I'm going to deploy another probe."

As he steps out and looks back to the southwest, he sees the funnel receding into the clouds directly above. "Damn luck," he curses, and presses another probe into the gravel, hoping that what weak vorticity remains will find his instrument. It has been an entirely frustrating intercept. The vortex finally approaches a single passable road, and it's already roping out.

The weight of yet another near miss settles heavily on Tim's shoulders. Another season may have just come to an unceremonious end. The weather pattern in the days and weeks ahead shows all the signs of settling into summer doldrums. He knows it is entirely likely that Nat-Geo will pull its funding. He may have no choice but to forge ahead on his own next year, again without financial support. How much longer can he make this work on a shoestring? The damn thing was so close you could *smell* the ground-up vegetation, you could *hear* the roar. Tim begins collecting his turtles, a morose expression on his face.

Gene Rhoden, the NatGeo guide, pipes up, "Tim, I see a golden color on the horizon."

Tim turns and gazes out to the east at the trailing edge of a thunderstorm catching the mellow light of the setting sun. The clouds are painted with the bright watercolor strokes the plains are famous for, but from this far out he can't judge the storm's strength. Tim ducks into the minivan and consults weatherTAP, a streaming radar service, on his salvaged cathode-tube-ray monitor. Suddenly, the fatigue dispels. The storm structure on the screen looks vigorous.

The day isn't over yet.

Tim throws the mud-daubed minivan into gear and tears off down the road to collect the rest of his probes. Then he hits the straight-shot pavement of Highway 14 and pushes the Caravan's six cylinders to some ninety miles per hour.

As they pass into the shadow of this new storm's anvil, the cab is filled with the vicious, singing hiss of wind-driven rain against glass. The sunset's warm apricot glow is replaced by dusk, the ambient light filtering through the clouds now sourceless and cool. The minivan approaches a low rise and a copse of cottonwoods, beyond which they are driving into the blind. As they pass beyond the trees and onto the table-flat tracts of soybeans and corn, the rain slackens, the sight lines clear, and the occupants of the minivan fall momentarily silent. The rain-soaked windshield is a phantasmagoria of liquid shapes, but there is no mistaking the profile before them.

"Wedge tornado on the ground," Tim says. "Oh, my God. It's *huge*."

"We gonna deploy on that thing?" asks Porter, his voice betraying more than a little trepidation.

"Damn right."

They approach from the west down Highway 14, the main route between Huron and Manchester. The tornado is half a mile to the south of the road and moving steadily northeast, refracting sunlight like a prism. One moment the mile-wide funnel is the color of sand. The next, it is smoke, ash, sod. Tim slows up, pulling into the oncoming lane. His distance narrows to hundreds of yards, but the approach is all wrong. There is the intuitive trimming along the margins of safety, and then there is the bet whose odds are unknown. From here, Tim can't discern the tornado's heading or ground speed with any certainty. This isn't the weakening Stratford twister. This is unlike anything he's ever seen. The tornado before them is the giant of plains legend, the breed a chaser may see once in his life. Even so, he won't chance a slapdash deployment. "I'm sorry, guys," Tim says. "This is too close for me. I'm not going in there."

On the radio, they hear the incongruously cheery sound of a plink-ing mandolin. Tim studies the DeLorme map and reconsiders his op-tions. There is another way, though the risk is still high. He'll have to leave the sure footing of pavement for a gamble on gravel. He drives thirty feet up the road and takes the next left, leaving the highway. "We've got some gridded roads," he reasons. "I'm going to go north." He'll use 424th Avenue, a dirt farm lane, to get ahead.

But before he drives much farther, Tim slows. Through the passen-ger window, no more than a third of a mile out, he sees the hamlet of Manchester, a huddle of oak, cottonwood, and whitewashed two-story farmhouses surrounded by wheat fields, the seed heads wicking gold in the sun. The minivan rolls to a stop.

"It's going north," Porter says. Neither speaks for a moment.

"It's going to take the town," Tim replies.

They watch the pretty old houses, the barns, the constructs of men standing pitiful and small in the growing shadow. First, a power pole leans and falls. A barn cants over, then its roof sails away. In mil-liseconds, the rest of the structure follows. A cottonwood, some one hundred feet tall, that has given shade to generations is flung to the earth. Now the tornado comes to the closest house. It isn't the roof that fails first. The entire two stories of it buckle so quickly as to be nearly imperceptible. The steeply pitched roof comes to rest on the ground. Then it is lofted several hundred feet into the sky. They hear none of the crack of splintering lumber, just the toneless, high fre-quency of white water, omnidirectional and immense. The funnel fills with white drywall, shingles, shredded pieces of insulation, large tree branches. They hang suspended, glittering in the sun. The destruction of Manchester—established with its own post office shortly after South Dakota's statehood a century before—takes only seconds.

As Tim and Porter resume the chase and gain distance, they glimpse the storm's totality playing out over a span of miles. The clouds are drawn to the core like water to a sink drain, then pulled into the tornado and centrifuged out. It is a sight into which one could lose oneself, but there is no time to linger. This road extends in a straight line to the north, but it won't intersect with the tornado. The storm is still bearing away from them to the northeast. That means Tim has to take the upcoming right, eastbound on 206th Street, and haul ass in front of the wedge over to 425th, the next north-south road. If he leaves enough distance between them and the tornado, he can drop the probe and turn north onto 425th before it's too late.

In other words, they are about to enter a race they can't afford to lose. One flat tire, one spinout in the wrong place, at the wrong time, and they're done for. But Tim may never get another chance like this— not this season, maybe not ever. He studies the map. "There's another crossroads in about a mile. That's where we're headed to. We're gonna be sitting in rain for a little bit."

Tim hits the gas, and the race of his life begins.

As soon as they enter the rain curtains surrounding the tornado, the tires begin swimming over cake-batter mud. Even by Tim's standards, this is madness. The road is a boggy mess. While he struggles to keep the minivan from sliding into the ditch, the tornado outline that had been so crisp begins to fray in the rain. There's a corollary of rule number one: keep your distance, *especially* from rain-wrapped tornadoes.

"We're losing visibility of it. Are you going to deploy in the rain?" Porter asks in disbelief. This is like playing chicken with a train they can't see. The minivan is fishtailing now; the road is getting worse.

"Yep," Tim replies.

"You don't have much time, Tim. Do *not* get stuck."

Tim doesn't respond. The rain is easing up now. The ragged, dusty wall bearing down looks different; it has narrowed, hardened, its outlines growing laminar, almost glassy. At least they can see it again. Tim hits 206th and hooks the right turn east.

"This is too dangerous," Porter says.

"It's all right," Tim says evenly. "It's still about half a mile out."

As the tornado draws nearer, its smaller features—aspects of its

ground-level flow, where radial winds hurtle into the vortex and suddenly turn and spiral vertically—come into view. They see the dust rise
and wrap around its back side. Black sod hangs in suspension around
the funnel like swarming flies.

They are closing in on the intersection with 425th Avenue, and so
is the beast. The road ahead, to Tim's relief, is paved. They'll be able
to make their escape to the north in a hurry, staking their lives on the
proposition that the tornado will continue to bear away to the northeast.

"Here we go," Tim says.

He can see his deployment site, just before the turn.

It is seconds away.

Now.

He slides to a stop near a dense row of poplar, behind which is a
farmhouse neither of them notice. The minivan's sliding door slams
open, and Tim removes a probe.

"Tim." Porter's voice takes on urgency as he watches dark shapes
translate across the tornado's face. "We don't have time. We don't have
time. *Seriously.*"

They are being tailed by Peter, the NatGeo photographer, whose
guides are growing deeply uncomfortable. Tim is so focused on his
goal, it seems to Gene Rhoden, that he has blinders on. Floating debris tumbles only a couple of hundred yards out, pieces of trees and
the next farmhouse over. The smaller, granulated detritus has already
begun to flutter just beyond Porter's window, like large snowflakes.

Tim is aware only of the tornado and his turtle. He switches on
the data recorder and eyes the storm's trajectory. He plants his probe
firmly onto the loose gravel.

Then he's sprinting, kicking up mud in his wake. He flings himself
back into the minivan and floors it. They swing left onto the paved
road, and through the windows they hear the white noise. Only now it
sounds like the thundering of Niagara Falls. As they pass the farmstead,
they can barely see the house through the rows of lush trees and the
minivan's mud-spattered windows.

"My God, Tim." Porter exhales. "I hope you got it. Did you get it turned on?"

"Yeah." The tornado still bears down.

"It is right on Carsten's butt," Porter says.

Something, maybe insulation, falls into the road ahead. Behind them, they see only the headlights of the Rhodens' vehicle shining against darkness. Porter watches closely through the rear window. "It just went through that house," he says.

After gaining some distance, Tim begins braking. The tornado is less than a mile behind them, but they've earned some breathing room. "Okay. I want to stop and look at it." Tim steps out onto the wet asphalt and stares up at the thing he has just outrun, now a long, sinuous trunk boring into the prairie. It leaves the farmstead and crosses the road near the probe, ejecting the trappings of a life into a field of young corn. There is something about the way the ephemeral vortices within the circulation lash over the ground, like ripples on water. They can see soil moving radially to the center and propelling skyward at speeds that seem to violate natural law.

"Listen to it," Porter shouts, awestruck.

Tim plants a new probe he has outfitted with cameras. His white socks are pulled high over mud-slathered calves, and his dark hair is wet and matted against his forehead. The tornado swings to the east, into the corn, and seems to hover in place. Meanwhile, Peter, the NatGeo photographer, deploys his own photographic probe, called Tin Man.

Suddenly the tornado curves northwest, toward the caravan. Tim, Porter, Peter, and the Rhodens pile into their vehicles, laughing and hooting, and begin to flee. They speed north for less than a mile before slowing up again. Gradually the vortex constricts, transitioning into what Tim describes as a laminar tube, narrow and translucent, like spider silk lit up by the setting sun. Another metamorphosis, and the tube is replaced by a thin, tattered strand that hangs and twists in the air like frayed cable. Slowly, it all dissolves into blue sky, leaving an amputated skirt of dirt and rain on the ground to whip and gust what remains of the low-level rotation. That, too, then begins to fade. The most incredible tornado Tim has ever witnessed—the true giant for which he has searched more than a decade—has finally breathed its last.

"That probe you put down, before the camera?" Porter says. "That took a direct hit."

Pink insulation flutters down onto the cornfields. "Good." Tim stows another probe into its plywood drawer. "I hope it's on the ground."

————

They retrace their steps, scanning the road and ditches ahead for Tim's turtles when they come to a freshly destroyed farmstead. Tim parks nearby and begins to pick his way through the wreckage. Debarked trees, blasted with topsoil, are piled atop one another like kindling. He passes a barren concrete slab. A mud-coated basset hound shakes itself off in the road. There is no sign of its owners. A few other chasers and locals have arrived. One of them calls to the animal: "Come here, doggie. We need to put you on a leash."

The dog rolls over onto its back, tail thumping between its legs.

"Yes, you're a *good* dog."

The Kingsbury County sheriff addresses Tim's group. "Are you people the storm chasers?"

"Yeah," Tim says. The air is heavy with the scent of propane.

"I don't want people all over out here."

"I completely understand. Is everybody all right here?"

"We've got a propane tank that's leaking."

Far off on the horizon, Porter sees the slender profile of another tornado. "It's hosin' again."

The sheriff, Charlie Smith, squints into the distance and his eyes widen. "Oh, Christ! That's over by my house."

"Was anybody home?" Porter asks, turning to the farmstead.

"I don't know," Smith says. "I don't know. I can't . . . I hope Harold got out."

The sheriff goes shuffling off, looking lost, his utility belt clinking in the ruin of a home belonging to a man he can't find. He stands at the edge of the exposed cellar and peers down at the wooden beams, planks, and cinder blocks that have collapsed into it. "Harold?" the sheriff calls.

Tim steps warily onto the rubble and cranes his neck to look into its darkened interstices, confronted with his tornado's work. No signs of life; no signs of death either. He seems to lose himself as he stares into the shadows of the cellar. Then he remembers his own role. He turns away and begins stalking toward the minivan. "I gotta go south and get my probe."

He passes a flattened farm truck, the cab shorn flush with the dash panel. Tree-root boluses the size of tractor tires are upturned and naked. He looks back toward the house and sees the funnel a few miles out, deep blue and lancing diagonally at the fields. He hears birdsong, and the mournful bawling of mortally wounded cattle. Tim and Porter step through a mosaic of lumber, all arrayed in the same direction, the wind field's fingerprint left on the earth. "This looks like F4, F5 damage to me," Tim says.

They climb into the minivan and drive away, in search of the turtle that Porter saw get hit. For all Tim knows, it has been carried off or smashed by debris.

Porter's wife—Kathy's sister—picks this moment to check in. "Hey, guys, is this a good time?" she asks cheerily over the vehicle's speakers.

"No!" the two shout in unison, and hurriedly end the call, with a promise to phone later.

As they scan the road ahead for signs of the bright cone, they find no trace of it. Back the other way, NatGeo's Tin Man is wedged in the mud some 500 yards from where it was deployed, its glass ports smashed, the camera inside ruined. After everything, perhaps HITPR hasn't survived, either, its shape unable to resist the strength of a giant. Perhaps these years have been wasted.

Tim pauses and consults the DeLorme road map—he remembers the probe had been geotagged upon deployment. According to the map, the turtle is in the *other* direction. He passed it? Tim whips the minivan around and returns to the farmstead. The row of poplar he'd deployed next to is gone, and all that had been green is now the same monochrome mud gray. Harold's house is the very farmstead he'd deployed next to. He simply had not recognized the area in its current desolation.

"There it is," Porter yells. "It's still there!"

The minivan comes skidding to a halt, and Tim strides purposefully toward the turtle, a bewildered look in his eyes as he glances over his shoulder at Porter. The probe is in the path, surrounded by a degree of damage that could only have been caused by the tornado core. History has been made on a dirt road in South Dakota.

Tim bends and examines HITPR, then grins, pointing. "There it is. I'm not going to touch it. I need to get some shots."

Peter, the National Geographic photographer, comes running up, breathlessly repeating, *"This is amazing! This is amazing! This is amazing!"*

Porter informs Peter that he is standing on fallen power lines, mercifully carrying no current. Tim begins to recount the event, as though he were a quarterback re-creating a bootleg play. His words come fast as he acts out the scene for the photographer. "That tornado was right there, and Pat says, 'We don't have time, you gotta go.' And I said, 'I'm gonna put it right here.' I reached out and slammed it down, made the corner, and we took off."

Without another word, Tim heads back to the minivan, his sneakers making sucking sounds in the mud. He returns with a camera and snaps a few photos. "Now, the sixty-dollar question: Is the little light still blinking in there?"

He grips the turtle's soil-spackled rim, lifts it, and pauses for a moment, regarding an unbelievable sight. Underneath the HITPR is a perfect circle of dry gravel road. All around it, the gravel has been scoured away by the tornado, leaving nothing but the rain-soaked roadbed. But beneath the device, the road is pristine. A small red light flashes.

"Yes, it is!" Tim shouts. "Look! The gravel is still here, right underneath it!"

He kneels in the mud, his probe balanced on its rim in front of the indisputable evidence that it hadn't budged an inch. A few yards away, on the other hand, Harold's house has been shredded and cast into an adjacent field. Some eight inches from the device is a deep gash in the road. Something big and tumbling had come very close to destroying the turtle. "These are F4, F5 winds," Tim says.

"I think it's F3 from what I see," Rhoden ventures. "That house is not anchored whatsoever. Certainly no higher than F4. Could be stronger, but no stronger than an F4."

"Well"—Tim grins, tapping his fingernails against the mild-steel skin—"I can certainly tell you."

———

That night, in a Huron, South Dakota, motel, Tim plugs his laptop into the turtle's data card. He is surrounded by the Rhodens, Porter, and Peter. They are all exhausted, wet and filthy, but the room hums with excitement.

In the lamplight of the shabby room, Tim's eyes dart over the numbers registering the pressure drop HITPR has recorded. "I've heard estimates of seventy, maybe eighty millibars." he says, looking up into Gene Rhoden's camera. "But not one hundred. Not *one hundred* millibars."

What Tim's turtle has measured is, as he'll come to describe it, the eardrum-shattering equivalent of stepping into an elevator and launching 4,000 feet up in ten seconds. Today, HITPR has collected data that never existed before now—it was always an estimate, a theory, an assumption, a blank space in the equation. Tim just filled that space with something real and precise. The pressure drop the turtle logged, using technology Tim had developed on the test range years before, is the steepest pressure deficit on record, confirming what scientists believed but could never prove. With a back-of-the-envelope calculation, and a few assumptions about the vortex structure, the probe's pressure ports registered winds in excess of two hundred miles per hour, well within the F4 range.

Tim used to be just a tinkerer with a high school diploma. Now he's the man with the impossible data, the first to step forth into an untouched—and supposedly untouchable—frontier.

Even now, in the giddy afterglow, when he should be cracking a Coors and toasting a watershed moment in scientific history, Tim speaks as though he is matching wits with his detractors and doubters.

"Certainly, you can open some of this up for speculation," he says. "Now, even if we were to hedge that . . . even if you average these small points down here, you're still going to average out close to one hundred [millibars]."

"It's amazing," Rhoden says.

"This is absolutely amazing."

"You've done it, Tim."

He looks up at them from the motel desk and his eyes gleam. "I think everybody did well today. It's fantastic."

Tim stays up late talking, telling and retelling the story of the day. No one in the room likely gives much thought to the danger they had courted today. Not now. They're all too wired. Tim dreams about what the numbers in his computer could mean—to forecasters, to structural engineers, and to folks like Harold, whose farmhouse stood for only an instant next to the turtle. Tim can tell them how fast the wind was at the surface. He can tell them about how the temperature steadily fell and the humidity increased as the tornado passed over HITPR. At last, he can tell them about the structure of the vortex. For researchers such as Bill Gallus, aiming to understand and protect against the near-surface winds of the tornado, these first-of-their-kind measurements are priceless.

As of tonight, the tornado has kept its secrets long enough.

PART TWO

CHAPTER ELEVEN

DOUBLING DOWN

THE YOUNG BOY in his bedroom, tinkering with old radios; the newbie chaser on the Colorado plains, casing an F0; the engineer at the test range, quantifying the raw violence of missiles and bombs. At every stage it was beyond a long shot to imagine this young man swooping down into the heart of the tornado, succeeding where decades of scientists had failed. Yet at each stage, he was gaining the skills he needed.

Now he's become the man who accomplished meteorology's equivalent to the moon landing.

In the aftermath, Tim is the toast of the weather world. *National Geographic* puts him on the cover, with a full story detailing his exploits. He travels to Chicago to appear on *The Oprah Winfrey Show*. Next he's on the CNN set, sitting across from Soledad O'Brien. He's wearing a clean pair of khakis and a pressed Hawaiian-print button-down. His dark, coffee-colored hair is smooth and neatly parted, for once not windblown.

For Tim, the most important event is the Conference on Severe Local Storms, the atmospheric-science equivalent to the Academy Awards, which he and Julian Lee attend over a year after Manchester.

The conference is a kind of coming-out party for the bleeding edge of severe-weather research, attended by the most august atmospheric scientists in the world. Tim and Lee unveil the first data of its kind—pressure, temperature, and humidity measurements from inside the core of an F4. Tim's probe has proven so capable that in its data log one can see even the brief deviation in pressure caused by Harold's farmhouse, as the tornado chewed through it and across the probe.

Their reception at the conference is resounding; the conclave of scientists is completely won over by Tim's findings. "It's extremely valuable," says Dr. Al Bedard, the senior NOAA scientist behind TOTO. "My perspective is that this was the first time a direct measurement was ever made." After years of cool skepticism, Tim basks in recognition and praise. His spiritual predecessors—Bedard, Howie Bluestein, and Joe Golden, the very men who inspired him from the beginning—offer sincere encouragements for him to continue his work.

Tim wants nothing more than to do just that. With the Manchester deployment, he has broken through the psychic hurdle that had dogged him and every severe-weather researcher from the very start. He has snatched fire from the god of the plains. And now, despite the clear dangers he faced in Manchester, South Dakota, and Stratford, Texas, his success only compels him to continue pushing forward. He bears the torch, and he can't fathom quitting the odds game Anton Seimon left back in Stratford.

Tim can return to the plains and do it again, he thinks. He can do it better. As long as he keeps the tornado in sight and establishes a reasonable buffer, the odds swing in his favor. He can capture an F5. He can invent new probes and collaborate with new scientists to gather more and more-complete data. Like the mountaineer who finally conquers an untouchable summit, Tim's eye is already on the next peak.

But the path forward is not so clear now. The big breakthrough has been made, and the next set of goals is different. Repetition, precision, and control are the keys to turning a daring exploit into sound science. Now that he's proven what's possible, he must leave behind his lone-wolf ways and collaborate. There are visionary scientists who can bend his ingenuity and chasing skill toward the field's most pressing questions: What is the signal, the mechanism in the atmosphere, that distinguishes a standard supercell from one that creates a twister? What dread signs mark those unforgettable days when tornadoes measure a mile across and scorch the prairie for hours at a time? These mysteries and others have bedeviled mankind since before Tim was born. But if he can repeat his core deployment and collaborate with others studying different parts of the storm, they finally have a new and long-sought-after tool to crack open these questions after years of stasis.

The most vexing question for Tim, though, is whether what got him here can sustain him, or whether he will need to change. His goals expand after Manchester, as he steps onto the scientific main stage. His drive will keep intensifying in turn, and his determination will continue drawing him closer toward the storm. But as much as he is galvanized by recognition from luminaries like Bedard and Bluestein, it's also true that he has long been fueled by his role as an outsider. He is brilliant but has long avoided the traditional route. He is confident and a strong leader, but he doesn't like taking second billing. He admires pedigreed scientists, but he can be thorny and self-conscious in their presence. He works hardest when someone tells him that something *can't* be done.

Now that he has made his name, can he fold himself into the role that the next scientific advance will require?

———

The first hint of an answer comes nine months after Manchester, when Tim arrives at the Iowa State lab of Professor Bill Gallus, the scientist who had written him off as a "yahoo." Gallus had cringed initially at the thought of collaborating with Tim: He's a chaser, not a scientist. But Gallus's engineers have continued to hound him about the blind spot in his simulator. "Could this be the guy who has the data we need?" they ask. As news of Tim's incredible measurement races through the research community, eventually Gallus gives in. What other choice does he have? There's only one ground-level data set from the core of a violent twister in existence, and it belongs to Samaras.

As Tim walks into Gallus's simulator facility in Ames, Iowa, the professor is nervous. He's preemptively embarrassed for Tim, worried that this hobbyist will be out of his depth among the engineers and the arcane language of aerospace and fluid dynamics. *How is this going to work?* Gallus wonders. Alternatively, what if Tim sees the simulator as a "hokey little toy" in comparison to the battle-tested turtle? In either case, this meeting could be a disaster.

But once again, Gallus has underestimated Tim. Far from floundering, he is positively bursting with insight. To the shock of Gallus's lab, Tim *is* an engineer. He speaks the language, and what's more, he's entranced by Iowa State's mighty machine.

The first time Gallus cranks up the simulator, the look on Tim's face is one of utter enchantment, "like a kid in a candy store." He marvels at how much the vortex resembles the Manchester monster. The way the dry-ice clouds feed into the vortex at lower levels, it's as if he's been transported back to a dirt road on the South Dakota plains.

"Can I put my probe in the path of your tornado?" he asks.

The engineers are chagrined. The minor winds produced by their artificial twister might register as only the most meager blip on HITPR. But Tim insists. He hauls out a turtle and places it onto the platform beneath the simulator. As the fan roars to life and the vaporous vortex translates over the loud-orange cone, Tim is ecstatic. The pressure fall is minuscule next to that of Manchester—about three millibars compared to one hundred—but its contours are eerily similar.

Over the course of the meeting Gallus's concerns about Tim's competency are thoroughly put to rest. Both come to appreciate the other's strengths—and in the presence of Iowa State's undulating dry-ice tornado, a partnership is forged. Gallus sets to writing a grant proposal to NOAA to fund Tim's probe-fielding attempts. In exchange, Tim will feed Gallus more raw data from real-life tornadoes. Tim's Manchester data set will allow Gallus to finalize his simulator, and in the coming years it will be used to better understand the complex interactions between structures and tornadic wind fields. As Tim gathers further data, it will serve as the foundation for numerical models that explain the anatomy of all manner of vortices chasers see in the field.

Tim's agreement with Gallus is promising, but in some ways it also underscores the handicap of Tim's background. It is unlikely Tim will ever secure a federal research grant on his own. A National Science Foundation reviewer would "shoot him down," Gallus says, based not on the relative strength of his ideas and abilities, but on the weakness of his academic pedigree. It's a fateful truth about Tim's new position. The closer one gets to the exclusive club at the top of the field, the more one is judged, as Julian Lee puts it, by "the number of letters you have after your name." Tim has little choice but to rely on others for funding.

Yet even in academia, Gallus is facing limitations of his own: the pool of money for tornado research is shallow these days, and there are bigger fish ahead of him. Between what Gallus can provide and a new round of funding from NatGeo, Tim's expenses will be largely covered for the upcoming season, but he'll be working for free. Paid vacation from ARA will only get him so far before company policy requires that he take an unpaid leave of absence.

Tim wants more than just to scrape by post-Manchester. It's not just about the thrill of the chase anymore. He wants to build a base of knowledge that can leapfrog structural engineers forward, that can tamp down the death and destruction caused by tornadoes. For his mighty ambitions, he will need a high-profile partner, one with resources. Look at what he's done without them. *Imagine what we could do together*, he wants to shout. Tim is already making his next moves. He is working on a new probe—larger than HITPR but just as durable, containing seven video cameras that can record simultaneously. Tim calls it the media probe. Six lateral cameras peer through apertures from behind Lexan screens. The seventh is angled vertically. Not only could the probe provide an unparalleled glimpse of the tornado core, it could, by tracking debris across multiple lenses, yield an even more accurate estimate of the tornado core flow than Tim's turtle.

In the late spring of 2004, near Storm Lake, Iowa, Tim succeeds in placing a prototype media probe inside an F3 tornado. The camera peers into an inhospitable place in which seemingly every centimeter of air boils with grass and soil and even larger objects, including a corn bin. The sound of it is like the crowning of a forest fire.

Just a year after his one-of-a-kind data, he has landed another tornado core and gathered one-of-a-kind footage. Tim's successes are coming in quick succession now. His name is on the lips of every active storm researcher.

———

It isn't long before one of the field's biggest whales, Josh Wurman, founder of the Center for Severe Weather Research, reaches out to Tim. The two had worked together briefly to analyze the Stratford tornado a couple of years back, and Tim knows Wurman's reputation well. The senior scientist is one of the perennial players in tornado research, reliably on the receiving end any time research dollars are on the line. After earning his doctorate at MIT, Wurman constructed the first ground-based, mobile Doppler platform in 1994, enabling him to haul a radar as powerful as any research installation's out into the field, where it's needed most. His Doppler on Wheels (or DOW) and similar mobile Dopplers developed by Howie Bluestein and others revealed things no weather-service radar had ever successfully imaged before. Wurman, for instance, holds the record for the fastest tornado wind speed ever captured, at 301 miles per hour.

Yet as much as mobile Doppler has offered, Wurman knows all too well that new angles on the tornado are needed for the full picture to come together. Chief among those is the near-surface level. Here, Tim is not just the leader. He is, Wurman says, "the only serious player."

That's why Wurman approaches him prior to the 2005 season. Wurman is in the early planning stages for what will become the largest tornado-research project ever mounted. VORTEX2 will be the successor to the original project VORTEX (the Verification of the Origins of Rotation in Tornadoes EXperiment), the sprawling science expedition that operated for two seasons back in the midnineties. This one will be even more ambitious, a historic expedition that seeks to answer the fundamental questions that continue to confound the meteorological world into the twenty-first century.

Ground-based, mobile Doppler radar was embryonic during the first experiment; the technology has since improved dramatically. Wurman envisions a fleet of mesonets and mobile radars encircling the storm, gathering data from all angles and elevations. Crucially, he shares Tim's conviction that unless scientists understand the characteristics of the low-level wind flow in tornadoes, we'll be powerless to guard against them or to create houses and buildings that can stand in the storm. This will require an all-important set of probes rugged enough to survive the worst tornadoes imaginable.

Wurman tells Tim that he's the best man for designing and placing the mission's probes.

What Wurman is offering is exactly what Tim needs after Manchester: the opportunity to work hand in glove with top researchers; generous federal funding throughout; a mission whose ambitions are as large as his own; and a perfect chance to build upon his breakthrough. This is the next peak that Tim has been eyeing. It seems as though it should be an easy yes from him.

But as Tim and Wurman get down to talking, something strange happens. A rift forms.

More often than not, Tim is, in the words of Julian Lee, "reverential toward academics, perhaps because he himself did not have the opportunity to pursue a formal higher education path." But there's a flip side to this reverence—Tim's pride despite his lack of degrees. When his work is challenged by decorated figures, he can easily turn cagey and defensive. This doesn't mesh well with Wurman. The scientist is blunt, and his bedside manner has never been honey dipped.

Also, unbeknownst to Wurman, Tim is already primed to be wary. While the two had collaborated once in the past—coauthoring a conference paper that combined their data on the Stratford, Texas, tornado—Tim was irked by what he perceived as the minor billing his deployment received. As Tim saw things, Wurman treated the turtle as a curiosity whose only purpose was to validate the Doppler on Wheels. Tim wrote to a colleague, "He's used my data with little/no mention of my efforts, and used it as a tool to promote his efforts against the other mobile radars out there."

Wurman says Tim never expressed these feelings to him after the publication of their paper, but he confirms that he saw radar as a more important tool than the in-situ probe.

Tim must have sensed Wurman's perception of the pecking order, and his place in it. Their history left Tim feeling "stepped on," he writes in an email. Which he might be able to get past. But as Wurman gets into the weeds, he doesn't succeed in alleviating Tim's concerns. Wurman tells Tim that he isn't interested in HITPRs for VORTEX2; instead, he's looking for Tim to invent cheaper, mass-produced versions of the turtle that can measure wind speed directly. As an inventor himself, Wurman admires the pressure-logging turtle and the elegance of its instrument package. "The thing is," he says, pressure "is the measurement we least care about. We know the pressure is low in a tornado."

What we don't know, Wurman says, is how the tornado's strongest winds are distributed near the ground. To that end, he doesn't want to settle for deploying—and landing—just one or two probes per intercept. "These onesie measurements," he says, "tend not to be too useful." Instead, he wants a swarm of probes, to capture the strengthening and slackening wind profile of the storm. If Tim's HITPR is a Lamborghini, Wurman is asking him to assemble a fleet of wind-speed-measuring Chevrolets.

The conversation rubs Tim the wrong way. But the long-term benefits of working with Wurman are impossible to ignore. Despite his reservations, Tim submits a one-page letter of intent to join VORTEX2 on December 3, 2004. He outlines his potential contributions, including his HITPRs and the new media probe, which should be usable to measure wind speed through the right calculations—but he doesn't mention any new Chevy probes.

In January, he discovers that Wurman has submitted his own letter of intent, which also includes a planned instrument with cameras, what he calls a pod. This sets off alarm bells in Tim's mind. Right or wrong, he's convinced that Wurman's "video pod" is an attempt to outflank the media probe and to underbid Tim's own proposal. Tim fires off an email to a colleague: "Being that Wurman is on the [VORTEX2] board [one of six] he was able to look at my submitted one-pager, and responded with a one-pager of his own describing HIS version of massive video probes . . . undercutting my cost." Already suspicious, Tim is now fully spooked. "Did I mention that he was on the V-II board reviewing all the submitted one-pagers?"

For Tim, this is the final straw.

Wurman dismisses the idea that he has undercut Tim's proposal; he's simply filling in the gaps Tim wouldn't. For now, it ends up being a moot point anyway: VORTEX2 needs to be pushed back, as NOAA and the National Science Foundation decline to fund the expedition for the time being.

Regardless, Tim's position stiffens. He comes to reject Wurman's criticisms along with the allure of his offer. As Tim sees it, he has a formula that's been successful. Who else has gotten probes into a tornado? And it has worked with the devices he has in hand, without any higher-up ordering him around. He has been told before that his methods would fail, and the doubters were wrong. He has been told before that he won't be able to make it as an outsider. But that's *exactly* how he has gotten by.

Tim took a chance at Manchester. He left behind the smooth, easy pavement of the highway and picked up a muddy gravel road. He made a bet on his own ability, on his wiles, on his force of will. And it paid off.

Now, he decides to make the same bet again. If he's not fit for Wurman's project, to hell with it. Tim will take inspiration from VORTEX2 and adapt it to his own strengths. He will make his own team—agile, wily, with Tim guiding its movements and his own deployments. He'll create a rival to VORTEX2 and the club of elite meteorologists.

This is his gravel-road option. He decides he doesn't need to fold himself into anyone else's system. He'll head off the storm's greatest mysteries using a route that no one else would dare take.

Tim leaves the easy road and crosses over into the mud. He slams the gas and doesn't look back. The decision to spurn Wurman seems small at the time—but its consequences will feed and spiral.

CHAPTER TWELVE

A TEAM OF UPSTARTS

I N 2005, TIM starts bringing together his collaborators. He's looking for scientists whose mission complements his own, who are tackling questions he and his turtles can't address alone. Their combined
efforts will create a full picture of a tornado, inside the core and out. If
VORTEX2 can leverage repeat measurements of the complete tornadic environment to resolve questions that have long stumped researchers, so can Tim's mission. Throughout the spring, a first-rate scientific
team emerges, with a distinctly Samaras character: trim, dedicated,
tight-knit, and with a bit of a chip on its shoulder.

First and foremost, Tim has in mind a woman he met last year.
In April 2004, he had been scheduled to speak at NOAA's Northern
Plains Convective Workshop in Sioux Falls, South Dakota. During dinner after the first day's presentations, he ended up sitting next to Cathy
Finley, a kindred spirit who ran a tornado-research program with her
partner, a scientist named Bruce Lee. As she explained to Tim, their
data-gathering strategy was to sample the near-tornado environment
by surrounding the funnel with sedan-mounted weather stations called
mobile mesonets. Tim had listened raptly. Their team sounded a great
deal like his own efforts—nimble, independent, unafraid to venture

beneath the shadow of the mesocyclone. This was a woman after his own heart.

Her career trajectory had lately taken an unforeseen turn, she explained. She and Lee had recently resigned from the University of Northern Colorado, having grown weary of fighting the administration for every dollar the school's small meteorology department deserved. Since then, they'd been working in the private sector, with a company in Minnesota. Wind farms and solar arrays, as it turned out, were in desperate need of meteorological advice. Both she and Lee had been raised in the state, which made this a welcome homecoming as well.

Throughout, neither of them had given up on their first love: severe storms. Finley grew up in Benson, a small farming hamlet situated along the Chippewa River, not twenty miles from the South Dakota border. As a little girl, she had been frightened by thunder. But as she grew older, the thing she feared most came to fascinate her. It would ever more be a prime focus of her life.

As they chatted over dinner, Tim and Finley discovered they had much in common. Conducting hard research while earning a living in the private sector, they agreed, was a struggle without end. As talk turned to Manchester, they realized their paths had nearly crossed at Harold's farmstead. While Tim was pulling up to what remained of the house, Finley and Lee were just leaving. In fact, one of their students had nearly stepped on the turtle. At the close of the 2004 conference, Tim and Finley had bade each other a fond farewell.

Now, more than a year later, their conversation remains lodged in Tim's mind. In mid-2005, he decides to hunt down her contact information. He writes to ask whether she and Bruce Lee would be interested in a joint field operation. Perhaps, he suggests, they should start their own "mini-VORTEX project" during the 2006 season.

Collaborating with Lee and Finley is an ideal fit for Tim. Their missions are parallel and complementary, affording Tim all the benefits of working with scientists as astute as Wurman, but without sacrificing his autonomy. Furthermore, Julian Lee has now moved on from ARA, so Tim needs someone to do the complex math and draw scientific conclusions from his data.

Partnership makes perfect sense to Lee and Finley as well. In an intensely clubby field, where egos swell, here is a man they can work with. The researchers are tackling the storm on different scales, but they're pieces of the same puzzle. Tim is attempting to draw out the inner workings of the tornado and how it behaves at ground level. Lee and Finley are trying to figure out what *creates* the tornado. What sustains it? What kills it off? What supercharges it? The answers to these questions could one day aid forecasters, while Tim's work could help vortex modelers and structural engineers. Both prongs cover some of the most vexing mysteries left in storm science. Together, they form a complete package.

Like Tim's rotating cast of chasers, Lee and Finley's is a humble outfit. Their toolbox consists of three mobile mesonets—atmospheric-sensor racks that look as though someone tore the plumbing out of the kitchen sink and strapped it to the top of a sedan. The largest component is a doglegging section of PVC pipe that ingests air past a thermometer and hygrometer (used to measure relative humidity), before exhausting it out the other end with a fan. At the top of the rack, an anemometer angles into the wind like a propellered weather vane, tracking wind speed. The whole rig—like a giant tongue used to savor different characteristics in the air—is secured to a sturdy Yakima bike rack, mounted to the top of a Chevy Cobalt, favored by Lee and Finley because it's one of the cheapest sedans on the market. The mesonet racks, unlike HITPR, are hardly exemplars of cutting-edge engineering. They're composed of common research-grade weather gauges.

But it is not the ingenuity of the tool, in this instance, that makes or breaks a deployment. It's execution, timing, and focus; the right mesonet spacing and coordination can offer whole new dimensions of data on a tornado. They can't tackle whole storms like the research groups supplied with millions of federal dollars. Instead, they've taken on a narrower slice.

What fascinates Finley and Lee most is a second downdraft that forms when the prevailing winds, roughly four miles above the surface, slam into the rear flank of the mesocyclone and are forced violently downward. Known as the rear-flank downdraft (RFD), this current of wind hurtling toward the surface is both drier and warmer than a normal, rain-filled downdraft. The phenomenon—the focus of Finley's dissertation at Colorado State University—has drawn in both Finley and Lee because RFD surges seem to correlate with tornado formation, and they likewise seem to be associated with intensification. They could be the accelerant thrown onto the fire, transforming strong tornadoes into beasts like at Manchester. But they're still poorly understood.

"It turned out to be a really big issue in the field," Lee says. "In terms of cost-benefit ratio, we got a lot of benefit out of a small budget." Even so, they'll need many more chases, and many more data sets. Repetition is what it'll take to pin down whether the RFD is indeed what's driving formation or intensification. "People think it's easy to get these measurements, but it's actually quite difficult," Lee says. "That's why you have to do it for a lot of years and accumulate enough cases to make generalizations. You get snippets of data. Or the road network might preclude you from positioning where you need to be. The storm may not cooperate. It takes a long time to collect enough data to say something."

Tim sympathizes with their plight. Manchester was a groundbreaking achievement, but he too needs to build up a larger data set before his work will translate that into real progress for people who are caught in the path of the storm. The arrival of Lee and Finley into his life must feel like salvation. Not only do they understand, they are living his experience, his frustrations—they're fellow travelers along the same path.

Truth be told, what excites Finley most about partnering with Tim is the prospect of strapping a mesonet rack to the top of his chase vehicle. He ventures into regions of the storm that scare her and just about every other researcher in the field. "Nobody collects data there," Finley jokes, "unless it's by accident."

———

Bill Gallus, the Iowa State professor, is the next addition to the team. He and Tim have been collaborating on probe data successfully for over a year, and it's an easy decision for Gallus to stay on. He now agrees to supply Lee and Finley with apprentices and mesonet drivers, drawing on his ranks of eager graduate students.

The other permanent member of Tim's squadron is a chase partner named Carl Young, whose obsession might burn even more brightly than Tim's. A University of Nevada at Reno grad student, Young conducted the environmental analysis of the Manchester tornado and has lately become a fixture at Tim's side.

Both a talented forecaster and a hardened road warrior, he's the only chaser Tim has hunted with who can match his legendary stamina. They met at a meteorological conference in 2002, and it was Tim who encouraged Carl to focus his graduate studies on the near-surface tornado environment.

Carl was raised on the West Coast and had tried his hand at a number of odd jobs—an economics major turned insurance-claims agent turned wedding photographer. He even gave acting a shot, though his only significant appearance was a brief part in the 1997 film *Against the Law*, starring Richard Grieco and Nancy Allen. Carl's character was killed in the climactic scene, and it soon became clear that his acting career wasn't going to survive either.

His life rounded another corner in 1998, just shy of his thirtieth birthday, when he was involved in a terrifying car accident. The experience impressed upon Carl the fragility of his life and prompted him to leave California, heading east. As a boy, he had loved to watch the summer thunderstorms whip whitecaps on Lake Tahoe. And as he recovered from his injuries and began his life again, he had a newfound urge to see the real monsters out on the Great Plains. Over the course of two months he logged 25,000 miles, chasing storms up and down Tornado Alley, and as far afield as the East Coast. In the days before smartphones, he'd call his father and ask for updates from the Storm Prediction Center, or for the shape of the supercell on radar. His constant refrain was "Do you see a hook [echo]? Do you see a hook?"—the telltale sign of a supercell with tornadic potential.

Young returned to Lake Tahoe after that summer a changed man. Weather, he determined, would become the focus of his life. He enrolled in Lake Tahoe Community College to get his prerequisite courses out of the way, then was accepted to a master's atmospheric science program at the University of Nevada, Reno. His professors soon discovered their student was a fanatical storm chaser. Come May, he had his priorities, and he struggled mightily when final exams conflicted with the peak of tornado season.

Carl had no interest in being cooped up in a lab, it turned out. The real science was outside, to be seen and felt and smelled and heard. Storms left room for precious little else in his life. He taught a few meteorology classes at the community college in the off-season and served briefly as the program director for the environmental organization League to Save Lake Tahoe. The position, however, quickly became untenable due to his lengthy absences during tornado season.

In 2002, he was awarded a $15,000 Sierra Pacific Power Company fellowship to further his field research into severe storms and tornadoes. It was in this milieu of chasers and geeks that he met Tim Samaras. Perhaps obsession recognized obsession. From then on, seldom would one find Tim beneath a mesocyclone without Carl.

———

As Tim fills out the roster of his team, Anton Seimon, his old chase partner, is frankly relieved to see him prepare an expedition with a group of trained meteorologists and scientists. "Carl is a very good forecaster, and a clear thinker. The people he worked with before Carl, to put it bluntly, were his next-door neighbor and brother-in-law, or anybody willing to drive a car," Seimon says. "What business they had getting in front of tornadoes is an open question."

Between Carl Young, Cathy Finley, Bruce Lee, and Bill Gallus, Tim has a formidable reservoir of talent. If Lee and Finley can sample the rear-flank downdraft while Tim and Carl pierce the vortex with turtles and media probes, their team could pull off a scientific coup that vastly exceeds their modest number.

Tim has dreamed of collaborating with great minds, and now he has found his crew. The days of vanishing into the plains with nothing but a friend, a minivan, and a few turtles are over. It is decided: 2006 will constitute a sort of trial run, to ensure they have the right chemistry. Come 2007, the inaugural season will commence.

———

As new collaborators and chase companions drift into Tim's orbit, drawn by a shared worship of the sky, 2006 brings a connection of an altogether different sort—a surprise not even the savvy forecaster could see coming over the horizon.

On March 24, 2006, Tim watches a twenty-eight-year-old Iowan named Matt Winter enter the banquet hall of the West Des Moines Marriott. Winter and his mother thread through the crowd of chasers and meteorologists at the 10th Annual Severe Storms and Doppler Radar Conference, looking slightly lost. In this crowd, Tim can see that Winter, who works in compliance for an online job board, is a little out of place. What the young man knows of meteorology has been gleaned from a few books, mostly chaser memoirs and beginner's texts. He considers himself more of an armchair enthusiast, content to follow the exploits of his favorite chasers, and to track tornadic events from the safety of his computer or television. Tim is one of the chasers he's most enamored of.

Winter's mother, Sherry, had previously arranged with Tim for her son to attend the conference and meet him. Long before Tim had settled down with Kathy, he and Sherry had been sweethearts, though it has been decades since they last saw each other. After Tim had secured invitations for them, the young man reached out by email, just to introduce himself. Tim had asked if Winter was a chaser or a meteorologist, or if he had attended Iowa State. Winter explained that he wasn't a chaser, though he'd been fascinated by weather ever since he was six, when he watched as a massive storm's eighty-five-mile-per-hour winds blew down trees in his babysitter's neighborhood. Like Tim, Winter did not have a college degree. The young man had started a family. He wrote that he had three children, whom he was raising with his wife, Soun. The conversation never delved any deeper than that, but they resolved to meet in Des Moines.

At the conference, Tim has only a few minutes before his presentation, but he happily greets Winter and his mother. They make small talk, and Tim promises to give Winter a copy of *Driven by Passion*, a new compilation DVD of Tim's best tornado sightings. With that, he makes his way to the stage, and Winter and Sherry take their seats at a table reserved for them near the dais.

Over the next half hour, Tim rolls video clips, regales the audience with war stories, and flips through carefully composed slides of plains twisters he'd chased during the 2005 season. He is not generally known as a speaker who gets hung up, whose words are punctuated by long, uncomfortable silences. Yet, on several occasions, Tim grows quiet, and his gaze seems to settle on Winter. The young man shifts uncomfortably in his chair.

Sherry, Tim soon learns, has come to believe something he himself may have begun to realize during his speech, as he peered down from the lectern at Winter's face. The day after the conference, she tells her son, "The more you got wrapped up in weather and tornadoes, the more I suspected it." There is a strong possibility, she says, that Tim is his father.

Tim and Sherry had dated in 1977 when they were both practically kids, around twenty years old. They went on a road trip together to Montana at one point. Then, without warning, she had ended the relationship. She left Colorado "abruptly," as Tim describes their parting, and returned to Iowa, where she reconnected with the man Winter knows as Dad. Soon, she was pregnant with her first son, Matt. Given the timing, she had always known there was a chance Tim might be the father, but she had hoped, for the sake of her future husband and the life they were trying to make together, that Tim was not. "It was a tiny suspicion," she had explained to Winter. She had never told Tim until recently.

Not surprisingly, Winter takes the news hard. Tim is his hero, but he isn't sure whether he should be thrilled or furious. Soon thereafter, Tim gets in touch. He tells Winter that he intends to purchase a DNA test to find out once and for all whether Sherry's suspicions are true. The kit arrives in the mail after a few days. In Colorado and Iowa, both men run swabs over the insides of their cheeks and drop the samples in the mail. Less than two weeks later, the results are in: there is a 99.97 percent chance that Tim is his father.

In Colorado, Tim has been having his own series of surprising conversations with his wife and children, explaining the existence of a son he didn't know about until recently, by a woman he had dated before Kathy. Apart from being stunned, everyone seems to be taking the news pretty well. When Tim tries to explain this newfound son to his colleagues, he seems a little chagrined. The revelation flies in the face of the image he has cultivated—the squared-away exemplar of the meteorological world whose path kept to an unerring trajectory.

Tim composes a long email to Winter, explaining the results of the test and its methodology—perhaps to introduce some order to a development neither can control. He writes to say that he understands that Winter is a grown man with his own family: "The DNA test shows I'm your biological father, but I know that doesn't make me your dad. I know you have a dad. If you want to be friends, or if you want to sever ties, I will understand. I know the circumstances are very hard, but I want you to know that I'm proud and happy to find out that you are my son."

They continue to correspond, and before long Tim invites Winter, his wife, and children to Lakewood, to meet the "Samaras clan," as Tim refers to the extended family. That July, Winter arrives in Colorado with his wife and three children in tow. The young man looks like a bundle of jittery nerves in advance of his crash introduction to a whole new wing of the family. Tim's brother Jack has flown in from Savannah, Georgia, just to meet Winter; and Tim's other brother, Jim, will be driving up from Lone Tree, Colorado, with his family. When the Winter clan arrives at Tim's bungalow, the introductions are dizzying as they meet Samaras after Samaras after Samaras. But soon enough Winter seems to relax.

While Kathy tends to Winter's wife and children, Tim shows him around the shop where he builds his gadgets and probes. Tim holds forth about the enormous antenna in his backyard, which he uses as an amplifier for his ham radio, and to scan for signals from outer space. As they stroll around the bungalow, the two men must be searching each other for the their own genetic echoes. Winter doesn't have Tim's prominent Greek Albanian nose at first glance. But he did as a child, before a serious car accident forced him to get facial reconstructive surgery that permanently altered his profile. Beyond that, the subtle clues add up. Neither man is particularly tall. Both possess generous plumes of chest hair. There is something to the eyes as well, a certain crystalline intensity. They come to see it, that Winter is Tim's son.

The two move on to Tim's new chasing rig. Gone is the minivan after Manchester, replaced by a vehicle more suited to an adventurer's hair-raising exploits. The black GMC 4x4 pickup sports a throaty exhaust system and a hail-dimpled paint job. Winter notices there is a gaping hole in the camper top, opened up by a softball-size hailstone. He laughs. "Why haven't you fixed that?"

"It's a war wound," Tim replies proudly. "You don't repair those."

The next day, Tim takes Winter by Applied Research Associates, so he can see where Tim works. Little by little, he introduces his son to his world. The Samaras clan envelops the Winters as though they are their own. Winter and Paul trade *Star Wars* references with matching fluency. Tim reads stories to his new seven-year-old grandson, Nick, as the boy sits in his lap and pulls at his eyeglasses. Jack holds the tiny hands of Winter's year-and-a-half-old daughter, Haley, and helps the wobbling toddler walk across the living-room floor. Before long the kids are calling them Grandpa Tim and Grandma Kathy. That night, they order pizza, and Winter looks on in amazement as Tim and Jenny douse their slices in generous slicks of hot sauce. Winter notes that he has always seasoned his pizza the same way.

After three days, Tim and the others bid their visitors good-bye. Tim will see Winter and his family again next year for his fiftieth birthday, when the chaser will whip up a batch of his famed green-chili burritos. Tim speaks of consulting with a geneticist to find out whether a fascination with severe weather is a heritable trait. How else, Tim wonders, to explain Winter's lifelong curiosity in isolation from his biological father?

Thereafter, they'll see each other sporadically. When Tim swings through Des Moines for a conference—or finds himself lured by the promise of an Iowa storm—he'll often grab breakfast or lunch with Winter. The young man confesses to having difficulty forgiving his mother for keeping her suspicions from him all these years. Tim's perspective is both measured—*How could she have truly known?*—and hopeful: "Don't look at what could have been," he says. "We need to live in the now. We need to accept it."

While the time they've lost bothers Winter, Tim assures him that they have the rest of their lives to make up for it. Both men are still young, far from the twilight of their days. Yet, as Tim well knows, even the brightest day can curdle over, plunged deep into shadow by the advance of the storm.

CHAPTER THIRTEEN

TWISTEX TAKES THE GRAVEL ROAD

IN FEBRUARY 2007, after a quiet but successful season for Tim's new team, Dr. Josh Wurman tries again. He pulls Tim into a quiet room at the Radisson Hotel in Aurora, Colorado, during the 9th Annual National Storm Chaser Convention to deliver big news: VORTEX2 is back on. It has received the green light for two years of generous funding in 2009 and 2010. Nearly every field scientist in the country will be jockeying for a spot in VORTEX2 in the coming months. And Wurman still wants Tim to design the mission's fleet of wind-speed-measuring probes.

Wurman lays out for Tim all the advantages of joining the historic effort. And he is characteristically blunt about Tim's prospects should he decline to join the fold. "You're not going to get funding," he says. "How else are you going to get a real grant to do what you want from somebody other than TV?"

This is Tim's final chance to choose Wurman's path, to avail himself of every resource, every tool he could never afford. He'd be contributing to a mission that not only includes probes and mesonets, but radar trucks, weather-balloon launch vans, unmanned aerial drones, damage surveyors—a diversity of observing platforms that has never before been deployed on tornadic storms. But if he accepts, he'll be bumped back down to role-player status. It could be 2001 all over again—one head in a much larger herd, swept along wherever it leads.

As he weighs his options, there's another argument for Tim to consider, lurking just below the surface. Neither man mentions the prospect of death. But even for someone as skilled as Tim Samaras, there are storms out there that can't be understood with one's eyes alone. Mobile-radar technologies aren't just another way of looking at the storm, they can make operating near tornadoes a great deal safer.

Tim should be factoring this in.

Wurman himself is exhibit A, carrying with him a famous story of his own brush with a disastrous storm near Geary, Oklahoma—where his million-dollar machine became an unintentional probe with Wurman and his crew trapped inside.

It was on May 29, 2004, just west of Oklahoma City. Wurman and his DOW encountered a *multiple-vortex mesocyclone,* or MVMC—the term Wurman would later coin for a broad tornado containing one or several intense subvortices. He was having trouble figuring out the structure of this storm; it was like a mesocyclone on the ground, streaked with roving pockets of deadly wind. By the time he realized he had ventured too close, it was too late: DOW had entered the main circulation. His vehicle was engulfed by winds so violent that they wrenched a closed door from the frame of the stout truck. Terrifying as it was, this part of the storm was probably survivable. The main circulation appeared to be roughly a mile across; it was the subvortex, though, that worried him. It fluctuated in width from between 200 and 800 meters, raking the prairie in complex cycloidal loops that were difficult to decipher in real time. Only by using the radar images on his screen was Wurman able to get an edge. He called out directions to the driver and was able to evade the subvortex that held the tornado's strongest winds.

Without the DOW, Wurman would have been blind. The feed from the nearest weather-service radar might not have picked up the subvortex at all. And even if it could have resolved the structure, by the time the next radar update arrived—usually once every five minutes—it would have been too late. The day might easily have ended differently. "We had our choice of hazard because of the mobile radar data: bad or worse?"

If Tim joins Wurman, he will have the insurance of DOW's watchful and all-seeing eye, a crucial fallback tool if he ever strays too close to a storm. Without Wurman, Tim is stuck relying on five-minute radar scans. And then what?

Tim knows that mobile radar can offer an extra layer of certainty within a storm—he's had chaser friends who've made great use of it. But whether he factors this into his decision in Aurora, we can't know. What we do know is that Tim remains wary of Wurman, that Tim remembers all too well the agony of 2001, that he ascribes his success in 2003 to his independence, and that his heart is already with the team of obsessives he has strung together.

When he says no to VORTEX2, Wurman is shocked—but more so, disappointed. "Maybe he thought he was being bullied when we were all in the same room, telling him to do something different," Wurman says. His warnings—about Tim's funding and hardware—were well meant, from the scientist's perspective, even if they failed to land.

Both parties move on.

Wurman decides that it's time to undertake probe construction himself. He's not the engineering whiz that Tim is, but the in situ model is too important not to follow now that it's been proven. "I got in the Pod business reluctantly because Tim's design and strategy didn't seem likely to answer the scientific questions that needed answering," Wurman says. He's hoping that his Doppler on Wheels combined with new in situ pods will allow for a comprehensive, unparalleled map of the vortex.

Tim and his crew don't miss a beat. The 2006 test run with Bruce Lee and Cathy Finley performed well, even in a fairly uneventful season. The team formed a budding friendship and spirits were high. In their minds, Wurman and his convoy can take up the whole highway if they like; Tim's upstarts will stick to the gravel road.

On April 9, 2007, Tim sends an email to his partners to make the collaboration official. Every research mission should have a name, or at least a tortured acronym, so they decide to christen their creation.

> Bruce and I have been batting around a few terms to use for our fielding this year. This is what we've come up with to date:
> Tactical
> Weather
> Instrumented
> Sampling in/near
> Tornadoes
> Basically, "TWIST."

TWIST sounds a little truncated to Tim's ear, and he proposes a slight tweak:

> What do you think about adding the EX (for EXperiment) to arrive at:
> "TWISTEX"?

They all agree. To Wurman's VORTEX2, Tim now has the one and only TWISTEX.

———

The inaugural 2007 season provides an object lesson in the elusory nature of the twister. The first big tornado of the year will surely go down in chaser legend. On May 4, a 1.7-mile-wide EF5 effectively wipes Greensburg, Kansas, from the map—but TWISTEX is nowhere near the action. Lee and Finley's young mesonet drivers are still buried under final exams. For the remainder of the season, apart from a second EF5 in Manitoba, Canada—where the team doesn't have approval to operate—no other significant tornadoes touch down in the plains.

In a last-ditch effort, Tim applies for his own separate VORTEX2 grant, intending to operate independently but within the expedition's umbrella. But Wurman's instincts were right; NSF declines to fund Tim's proposal. Lee and Finley have standing invitations to join VORTEX2 as well. But after working with Tim, they aren't considering those offers too seriously. It sounds as though they would be expected to pay their own way. And like Tim, they're now convinced that they will be better off operating independently of VORTEX2.

At a subsequent American Meteorological Society conference, TWISTEX arrives like the plucky upstart on a bigger gang's turf. Bill Gallus, the team's academic partner, remains a little uneasy with the perception that they've gone rogue. The rivalry—one-sided though it may be—might as well have been lifted from the script of *Twister*. As Gallus says, they're "the little guys with really good ideas, who have the most experience with ground-based deployment." And, so far, "they get run over by the well-funded VORTEX2 planning people." But that's just the setup—who knows whether Wurman's armada or Tim's raw drive will triumph in the final act.

VORTEX2 is about to unleash millions of dollars' worth of gadgetry on plains storms. The fight might seem unfair, unwinnable on the surface—not even worthy of being called a fight. But if there's one thing Tim excels at, it's finding tornadoes. This is where the underdog thinks he can again defy the impossible odds. Because even without DOW to watch over him—even without the unprecedented array of tools at VORTEX2's disposal—all the equipment in the world means nothing if the scientists wielding them can't catch the right storm. That's largely the arc of storm science's history: researchers trying and repeatedly failing to locate the sky's most unpredictable phenomenon. If the members of TWISTEX have any small edge, it's that they all have reputations as road warriors and highly experienced chasers.

Atmospheric field research is a wide-open frontier, theirs for the claiming. There are only a few mesonet data sets from within close proximity to violent tornadoes in existence. Coincident mesonet and in situ probe data sets are even rarer still. TWISTEX has the opportunity to leverage its chasing skill and agility to drive tornado science into its next golden era.

Tim is betting that the natural law of inertia will prevail at some crucial moment in the 2009 and 2010 seasons of Wurman's experiment. Chasing is about being able to turn on a dime. But when a heavy object like VORTEX2 is in motion, it tends to stay in motion. When at rest, the armada will tend to stay at rest.

That's when TWISTEX will step up. The game is on.

CHAPTER FOURTEEN

QUINTER, KANSAS

O N MAY 23, 2008, not long after the annual kickoff of TWISTEX operations, the team is tracking a storm bearing down on a small Kansas town called Quinter. After a pair of unusually docile seasons, the project is off to a roaring start this year. Today is gearing up to be the team's first shot at a monster tornado. It will serve as an initial test of TWISTEX's cohesion and coordination under stressful, highly dangerous conditions. In equal measure, it's a test of the gamble Tim made by passing up a berth with VORTEX2.

From the lead car, known as M1, Bruce Lee and Cathy Finley choreograph the movements of the three mesonets. The convoy proceeds evenly spaced and in single file down straight farm-to-market roads, their roof-mounted mesonet stations cruising above the corn like shark fins.

If they're lucky, the stations will sample a cross section of the environment feeding the tornado. And if the weather gods truly smile upon TWISTEX today, a rear-flank downdraft surge will sweep across the cars, its characteristics captured by the dataloggers. Ideally, the tornado at the edge of the surge would pass over one or several of Tim's turtles and media probes. But first they'll have to contend with the radios that keep the team in fitful contact, limited Internet connectivity, roads that end abruptly, and storms that seem to resent the prodding of meteorological inquiry.

To Lee's and Finley's mild irritation, the list of obstacles also currently includes Tim and Carl, who have wandered off, gotten separated from the group, and are now irretrievably out of position. If the mesonets do manage to collect a data set this afternoon, they're unlikely to be bolstered by any probes. It isn't that they expect Tim to remain within eyesight; his black GMC is autonomous by design, able to venture far closer than the Chevy Cobalt and its four cylinders dare. Yet to Lee and Finley, their data set will be of little value if probe and mesonet don't deploy on the same tornado. At their best, each informs the data collected by the other. The full team's data is more valuable than any piece in isolation.

But M1 can't worry about that now.

The mesonets tack north along a sodden dirt road, tires leaving two-track depressions that quickly fill with standing water. To the northwest, a mesocyclone scrapes its belly across the prairie. As they approach a stand of trees, a thin, pearlescent vortex to the northeast terminates in a swirling wreath of dust. It crosses the road ahead, and a fuse or transformer farther down the line explodes in a cold, blue burst of light. When they near the source of the flash, the poles lean ominously overhead.

Within minutes the thin rope of a tornado swells into a textbook stovepipe, which they easily track over the flat cropland. M1 continues along at a leisurely pace, roughly twenty miles per hour, the optimal speed for gathering data. Apart from the fact that M3 had gotten turned around in Quinter and is lagging behind, M1 and M2 are in ideal formation. All that remains to be done is to keep good spacing and observe.

The storm, however, is only just baring its teeth. After M1 passes an electrical substation, Lee and Finley are blasted by a 108-mile-per-hour gust. It slams broadside into the sedan in what feels like an intense rear-flank downdraft. Suddenly the tornado they were following disappears. There it was one moment, such a terrible and beautiful thing to behold, tearing down power lines and ripping up crops. The next, it has vanished, like an apparition.

That probably isn't good, Lee thinks.

At this moment, M1 can be seen in a video taken by another chaser named Chris Collura, who is at times no more than fifty yards behind. The tornado is just to the northwest in one instant, and in a blink it's gone. All that can be heard is a low, moaning westerly wind. There's a hollow resonance to the sound, as if the empty fields themselves shape its register. What they hear next affirms that something has gone terribly wrong.

The low moaning jumps up the scale in Collura's video until it arrives at a pitch so high the camera's microphone defaults to stuttering silence. Lee and Finley look from the road ahead to the fields through the driver's window. What had been a single wedge off to the west has undergone a radical metamorphosis. This can't be the same tornado. Its borders no longer possess the clean, discrete curvature of a classic vortex. It looks as though the entire mesocyclone has fallen to earth. So wide is the funnel—if such a word even applies here—that it will not fit into the viewfinder of Collura's camera. As it roars northeast at more than forty miles per hour, half a mile separates Lee and Finley from a terrifying species of tornado they've never before seen. Collura angles his car into the wind. A broken window sprays the interior with glass. The dim afternoon darkens further, and all he can do is shout, "Son of a bitch! Fuck! My God. My God."

Ahead of M1, power poles quiver like tuning forks and suddenly pitch over. Black transmission lines lace the road. Lee and Finley career into the muddy ditch, tires spinning uselessly as the Cobalt slides to a stop. The tornado just beyond the window looks more like a sandstorm—its towering bluffs turning day into night. It's so big they can't tell whether it is moving toward them or away.

The only thing that's certain is that they can no longer move. M1 settles into the muck. All of the day's promise has evaporated into panic. For the first time since she was a little girl, cowering from Minnesota thunderstorms, Finley is afraid.

But even as the dark wall moves to within 800 yards, it is already hooking to the north, away from them. Lee and Finley can hardly believe what they have witnessed. What had been a moderately sized tornado one second became a monster in the next. Even before they have had a chance to look at the data, they know they've sampled a highly unusual rear-flank downdraft. The spike their temperature sensor has collected is the warmest they have yet observed, lending further credence to their theory: This particular RFD may have behaved like an engine's injector, dosing the tornado with jets of buoyant air. The result was explosive upscale growth—a fundamental phase change. More practically, it just threatened and spared their lives all within a few heartbeats; the RFD was what hurtled them into the ditch, and what steered the tornado away.

Finley and Lee's reprieve is momentary at best. New storms are building and will soon track their way, with M1 hopelessly mired in axle-deep mud. Before long, Tim's friend and mesonet operator Tony Laubach appears, driving M3, with Chris Karstens, an Iowa State grad student, in M2. Lee and Finley are further chagrined to see that a low-hanging power line has snagged Karstens's anemometer and torn it from the rack. On TWISTEX's first encounter with a high-end tornado, it is clear they have come up against something for which they were not prepared. Everyone feels a bit helpless, milling around the floundering Cobalt.

Then they see Tim and Carl maneuver around the downed power lines toward M1 in the 4x4. The two step out, grinning like kids. Carl gawps at the departing storm to the north. "Oh my gosh," he shouts. "That whole thing's a tornado!"

Tim lugs the winch hook and cable over to the Cobalt. "Man," he says, as he fastens the hook to the undercarriage of M1. "We got a great view of this amazing transformation from a stovepipe into that huge wedge. Looked like the whole meso dropped right down to the ground."

Lee and Finley, still shaken from the encounter, saw it, of course—and they hope never to see its like that close ever again. Tim strides back to the truck, activates the winch, and begins to pull M1 from the ditch. As the sedan lurches out of the mud, they hear a loud and disconcerting *bang*. The Cobalt is back on the road, but Tim has accidentally yanked one of the sedan's springs out of the control arm. As the sky continues to boil, M1 is crippled.

While Tim works, Carl monitors radar. The danger isn't imminent, but they need to get moving soon. There is another supercell headed their way, and it has a hook echo.

Lee and Finley are prepared to abandon M1 to its fate if it comes to that—but Tim has another plan. He retrieves a length of clothesline from the truck, loops it over the spring, and orders one of the grad students to stand on it. If he can fully compress and load the spring, he can fit it back into place. But none of them are heavy enough. Lee can almost envision the gears in Tim's analytical brain turning the problem over and over.

At last, Tim strikes on the solution. He jacks up the rear end of the Cobalt, places the coil beneath the frame, and lowers the vehicle, using the sedan's weight for compression. After lashing it with a slipknot, he shimmies beneath M1, fits the spring back into place, and cuts the cord with a pocketknife. He stands, wiping himself off, covered head to foot in viscous clay off the dirt road. The team stares from the newly mobile M1 to Tim in dumbstruck awe. "He has this presence about him," Karstens says. "He's very calm and collected." It seems Tim always has an answer, a fix.

The mesonets are finished for the day. They head for the nearest town. Tim and Carl, meanwhile, pile back into the truck and tear off toward Quinter. They're still in the hunt as their taillights recede beneath the shadow of the storm.

———

Minutes later, on the outskirts of town, Tim steps out of the truck and walks to the rear, shoulders stooped against the cool shock of rain. He pops the hatch to the camper top and lowers the tailgate. His instruments—the turtles, the media probe—are stowed top-down in cutouts on a sliding plywood deck. With a single practiced motion, he seizes HITPR's rim with his fingertips and lifts it over his head. He takes off scuttling toward the entrance of a park on the north side of town. The tornado is now approaching from the south. Behind him, Carl Young's camera sweeps over the outlines of a playscape, a few picnic tables, and an embroidery of small shade trees at the end of a narrow road. The camera frame centers on Tim, who balances the probe on its rim, activates the data recorder, and sets it down gingerly onto wet grass.

They jump back into the truck, and Tim navigates down the town's few streets, his ropy forearms slick, and his dark hair matted against his skull. Carl uses the interlude between deployments to check his laptop. "Ah, this storm has got incredible rotation," he says, noting the storm-relative velocity. "The chance for this to have a big tornado is high."

In the soft afternoon rain, nothing moves in Quinter, a cluster of corrugated aluminum buildings and wood-frame homes. The town's sirens are silent, and the windows of the houses show no signs of life. It looks as if everyone has picked up and moved on, the way warblers are known to depart days in advance of a bad storm. Tim pulls off road again next to an empty lot and begins unloading media probes and turtles onto the freshly mown grass. If anyone is looking from out of the dark windows, they must watch him with a mixture of curiosity and misgiving. He is drenched, winded, ducking through the rain, planting squat cones of uncertain purpose around town in inscrutable configurations. If any onlooker understood the meaning behind his actions, and what they foretell, they would surely flee.

Tim and Carl once more climb into the truck and drive to the western edge of Quinter. The wedge tornado has veered, and they watch it move against the limpid sunset. The outline in the distance now tracks toward the uninhabited fields to the north-northwest—away from town and the probes that lie in wait.

They give chase past rolling irrigators strung over the dead-level fields, looking for another deployment angle and finding none along the tornado's careering path. It resembles a swaying elephant's trunk, trawling the horizon. Then the tightly coiled circulation pulls apart and drifts indolently through the sky, like gray-black ink dispersing in water. The storm pulses and the vortex darts earthward again, boring down into the empty fields, kicking up a sod annulus that rises along the updraft. The loose drifts of dirt end up getting recycled, entrained in the downdraft and knitted into tubes of dense vorticity.

Tim knows he can't safely deploy on this tornado; it's erratic and he doesn't think they can get ahead of it with enough time to drop the turtles. "I don't know," he says, "this is getting pretty dangerous. Very, very dangerous. One slip and we're all dead." It's a feeling in his gut, intuited through experience, distance, and pace. He and Carl are out of the game, spectators, not participants, in the first big TWISTEX field operation.

He may have missed the strongest tornado the team has seen since its inception, but he is fine with that today. Tim will push—and has pushed—beyond what other chasers will tolerate to gather the data he needs. But there are clear limits. He's conscious of the danger the mesonets encountered earlier. And with nothing but dirt roads to maneuver on, he's not about to risk their lives on a twister tracing an oddball track.

He and Carl chase for the remainder of the afternoon for the pure enjoyment of it, until the storms contain nothing more than straight-line wind and rain. Then they head back to the southeast to rendezvous with Bruce, Cathy, Karstens, and the others.

———

A few years ago, Tim unveiled his first media-probe footage—gathered from a tornado near Storm Lake, Iowa—at the National Weather Association conference in Des Moines. His current mesonet driver Chris Karstens, an Iowa State undergraduate at the time, was sitting in the audience. Before the video began, the attendees were instructed to stow all phones and video cameras. These images were licensed to National Geographic, but the conference attendees were to be given a first glimpse.

Karstens felt privileged, as if he was taking part in something historic. The house lights dimmed, and a hush fell over the crowd. On the screen, Tim's nose hovered over the probe as he carried it down the road. Then a minute or so passed and the debris cloud edged into view. Rocks and twigs and farm equipment flew across the frame, and it all seemed so impossible, so thrilling. No one had ever before collected video and audio from *inside* a twister. This was landmark footage from Tim's first successful deployment of the media probe. "He was doing things nobody else was able to do," Karstens says.

At the end of the video, the audience stood and the room echoed with applause. How incredible would it be, Karstens wondered then, to work with Tim Samaras? To tackle the fundamental mysteries of the poorly understood tornado boundary layer? Karstens quickly brushed the thought aside. He had student loans to pay and was in the middle of his junior year and an internship at a local news station.

But during his senior year, Dr. Bill Gallus told Karstens about an opening in the graduate meteorology program. He was looking to recruit research-oriented students to analyze data collected by Tim Samaras, and to conduct field work with TWISTEX. Karstens leapt at the opportunity. Gallus had intended to cycle students in and out of TWISTEX in two-week intervals, but Lee and Finley found the young man to be a quick study and a dependable mesonet driver. Karstens was desperately needed behind the wheel of M2, and he was only too happy to oblige; he idolized Tim.

As the season has gotten under way, there have been things the young man has noticed about Tim that might sound inconsequential in any other context, but for which Karstens is deeply grateful. When the power converter on his mesonet blew out shortly before a deployment, Tim cheerily repaired it within seconds. It was bizarre, Karstens thought, that when a malfunction could cost them a data set, and everyone else was tense and irritable, Tim's presence was lidocaine on a raw nerve. The more time he spends around Tim, the more Karstens sees his unorthodox background as an asset, not a hindrance. Earlier, the radios were giving Lee and Finley fits, so Tim figured out how to amplify the range without cranking the wattage and shedding electrical interference into their sensors. There seems to be no end to the patches, upgrades, and repairs he can perform in the field.

Tim is also, in a very real sense, keeping the team on the road through the funding he secures. Over the last few seasons, over thousands of miles, Tim's fundraising has sustained TWISTEX, even if the money has never gone quite far enough. The team is something of an outlier in this way. The other research groups have either full backing from a university or a big grant from the National Science Foundation. But TWISTEX's existence depends on Tim's charisma and connections to attract benefactors such as National Geographic.

Despite this role, Tim is careful not to let his whim or ego drive the team. Karstens has come to admire the deft authority he wields over the mission. When the team discusses potential chase targets, Tim listens to everyone in turn; and once he's heard every voice, he makes a final call. "No one really ever questions it," Karstens says. "He makes sound, logical decisions, and people listen." The young man can sense a connectedness running through the group, a tightness that feels like family. Karstens notices that Tim considers each of them his responsibility. Even though the veteran chaser has grown accustomed to operating in proximities that once frightened him, the prospect of losing a team member to a tornado haunts him. "I don't think I could live with myself if anything happened to anybody on the project," Tim says.

The night after Quinter, this potential is on everyone's mind. After traveling back through plains towns in the viridescent half-light of decaying storms, they've gathered at an Applebee's in Hays, an hour's drive east of Quinter. Finley downs a few drinks to steady her nerves. It is sheer luck that the tornado turned to the north, and they know it. Even Laubach, known for gleefully punching through supercell hail cores, says he didn't want to get anywhere near the thing.

As is customary, they begin sharing videos of the chase among themselves, then with other chasers. From different locations, each lens captured some feature that was invisible to the rest. Storm chasers have practically commandeered Applebee's, the only place in town still open at this hour. Most of them know each other, and as the din of conversation fills the bar, breathless intercept tales are told and retold. The day could have been deadly, yet all around, video cameras and laptops proudly display footage and images—the closer and more terrifying, the better.

Finally, a man looms over the TWISTEX team, bellies up to the table, and places his laptop confidently at the center. Doug Kiesling looks like a nightclub bouncer, a towering man with a gift for weaving remarkable, expletive-laced tapestries about the storms he's seen. He has cultivated a reputation as the guy who always gets the killer shot, even if it means getting a little too close. He cues up his video of the day, and the group is stunned to see that he had been within a hundred yards or so of the Quinter tornado at its most vicious. On the screen, he's putting his car in reverse and backing away rapidly as rain and gravel blast the side of his vehicle. The stovepipe is gone, and a darkness in the west fills the camera frame. They hear him exclaiming, rather calmly, all things considered, "I'm about to get hit by a fuckin' wedge."

Some applaud his narrow escape, and everyone agrees he has a remarkable piece of chase footage on his hands. The Weather Channel has apparently purchased the video already. But what it captures sends a chill through Tim's team.

The old axiom holds that they are far more likely to get killed on the road, on the way to a storm, than in a tornado itself. And it's miraculously true: no chaser has yet died in a tornado. But chasers continue to get closer and closer, goaded by the prospect of social-media glory, YouTube notoriety, and the modest sums earned by selling video rights to TV news.

For TWISTEX, the evening wears, as Lee puts it, "two faces." Meteorologically, they witnessed an incredible event—a tornado that seemed to assume remarkable power and proportions nearly instantaneously. They managed to squeak away with a captivating data set. But a hundred-mile-per-hour gust drove Lee and Finley into a ditch; a weak tornadic circulation apparently ran over the top of M2 before the vehicle was struck by a power line. They feel their size in relation to the storm, and they're shaken to the core.

Tim is troubled not just by what he's seen today but by the scene around him—a giddy celebration of tempted fate. But even as he frets over the danger evident in clip after clip, the arc of his own history bends toward ever greater proximity to the lethal winds. The greatest intercept of his life was no less a nail-biter than Kiesling's, though Tim would like to believe that the risks he takes serve a higher calling. Years ago, he was inspired by men who hunted tornadoes for science. He's one of them now. But the truth about implacable nature is that it doesn't distinguish between those who approach for knowledge, and those who approach simply to see something beautiful, frightening, or titillating.

At this impromptu gathering of men and women who are drawn to tornadoes like moths to light, Tim finds himself considering the odds game Anton Seimon faced back in Stratford, Texas. How long will the luck hold? How long before storm chasing's darkest day finally comes? Who's it going to be?

The hour grows late, but the beer is still flowing, and the mood in the bar is joyful. Tim takes it all in. What he says next to a NatGeo journalist who has been shadowing Tim this season has the ring of inevitability, of the borrowed time storm chasers have been living on: "Someday," he says, "somebody's gonna get bit."

CHAPTER FIFTEEN

"YOU HAVE MY ONLY SON"

I N 2008, TWISTEX welcomes its youngest member to the team: nineteen-year-old Paul Timothy Samaras, a budding filmmaker and photographer. In his eyes, Tim is a giant. Fourteen years after his father first took him and Jenny to see a funnel cloud near Aurora, Paul wants to understand the nature of the work that has drawn Tim so far from home over so many years. If he can, in some small way, he'd like to become a part of Tim's world. The truth is, Paul isn't sure exactly what he wants to do with his life, and joining his dad on the hunt seems as good an answer as any.

Nearly a week after Quinter, he gets his first taste of the narrow escape. Near Tipton, Kansas, Paul pours from the truck alongside Tim and Carl, a video camera held at his chest, eyes fixed on the display monitor. He circles around to the probe deck at the rear, the frame bobbing as his feet pound over asphalt. His microphone registers the tap of rain against the camera, and the rasping of Velcro straps from the truck. Through his lens dance the dark silhouettes of working shoulders, a father seen through the eyes of his son. Beyond them, suspended mist and rain gather beneath a wall cloud like a load of cinders. Tim squats and places HITPR in the grass just off the side of the road. The frame moves on to Carl, who labors under the bulk of the much-heavier media probe.

The tornado is now just off the edge of the frame, a presence implied but unseen. Then comes Tim's urgent voice: "Let's go, man!" And all three of them appear now in the lens of Tim's invention. Through the ground-level perspective of the media probe, Tim is seen shoving the probe deck back into the truck bed. There is Paul behind him, still filming. He's lanky in a billowing black T-shirt and baggy jeans, his dark hair wild and blown by the rising wind, like his father's when he was young.

Roughly thirty seconds have elapsed since they stopped here. Carl's voice is tinged with panic: *"Let's go! It's coming!"*

Tim raises the tailgate but leaves the camper-shell window open and runs. Finally Paul lowers the camera and scrambles into the backseat. Rain skates over the road. The rear bumper dips as the front end lifts and the GMC's eight cylinders redline. The tornado core soon flares into view. Vegetation and debris appear as oscillating streaks. A faint multiple-vortex structure washes over the probes.

Paul would have seen the spectral merry-go-round suction vortices dancing over the fields toward them at roughly thirty-three miles per hour. He would have heard their waterfall rushing. He might even have caught the scent of threshed grass. This is his father's world, in the tarnished light under the clouds. The beating in Paul's chest when they make good their escape—this is his father's adrenal high. Next to it, the days are but interludes once the skies go quiet.

Later on, the media probe's footage astonishes Bruce Lee. "They had fifteen seconds on that one," he says. "They deploy, and we see them leave, and the tornado runs it over in *fifteen seconds*. That's how close they pushed it."

When they hook a U-turn after the tornado has passed and return to the deployment site, Tim kneels in the stinging rain and lifts the media probe onto its rim. "All recording except for one camera," he shouts over the wind, and hoists the hundred-pound device to his waist.

One of the two HITPRs has malfunctioned. But the other and the media probe log direct hits—Tim's first since the footage he gathered near Storm Lake, Iowa. The three are ecstatic and can hardly wait to plug in to the media probe to review images from within the whirlwind itself.

Paul has witnessed every second of the deployment. Nothing will stop him now from following his father onto the plains.

———

Paul, like Tim at that age, is searching for a path. As a boy, he had been so different from the young man he became. Back then, he was wild and loud and outgoing, an active kid who glided up and down the streets playing roller hockey. In his teens, that boy disappeared. "Something just flipped when he went to high school," his sister Jenny says, "like some switch just turned off."

Paul stopped playing sports. He spent hours alone in his room on the computer. He had friends who were like brothers, but he was no social butterfly. To his family's knowledge, Paul seldom if ever dated, and there had never been a steady girlfriend. Perhaps there was nothing more to this introversion than an escape from the petty cruelties of adolescence. "High school was hard for him," his mother says. It was easier not to stick his neck out.

After graduating, Paul enrolled in Red Rocks Community College. He took a few astronomy courses, but found that his appetite for school was as weak as his father's. Paul seemed adrift, like any number of young men his age, looking for his own way, trying on and discarding possible lives. At first, he was obsessed with animated films. He went so far as to get a menial job at the local movie theater so that when his shift was up, he could duck into a movie. He could usually be found watching the latest Pixar epic—*Wall-E* or *The Incredibles*—over and over. His mother or one of his sisters often joined him, but he could never stand to wait for their schedules to line up, so he'd go see them alone the first time.

Paul wondered whether he, too, might someday be able to create worlds of his own, as beautiful and melancholy as *Wall-E*'s. At his request, his parents bought him a new computer and enrolled him in an online course called Animation Mentor, founded by animators who had worked for Pixar and Industrial Light & Magic. Paul completed the course—in fact, he was quite talented—but he never sought work in the industry, never created those films. Doing so would have meant relocating to California, and Paul couldn't leave behind the only world he'd ever known. It would mean abandoning the best friends with whom he could debate the finer points of death metal and the video game *Halo*. It would now mean abandoning the storms and chasing with Tim.

Paul decides to remain in Colorado instead and continues to live in the Lakewood home that has sheltered him since he was born. Instead of animation, he shifts to photography. No one is certain precisely how he settled on the craft, but his mother suspects he watched the *National Geographic* magazine photographers who'd sometimes shadow his father. One day, Paul simply picked up Tim's DSLR and started shooting.

Photography makes sense for him. Though he is quiet and reserved, he absorbs everything around him with big watchful eyes, like his mother's. He takes photographs of tangerine sunsets, the deeply folded mountains to the west, and his charismatic tabby cat, Meekers. He handles videography and photography for both of his sisters' weddings, along with the birth of Amy's little girl.

His video camerawork, shaky at first, gradually smooths out. The frame now glides over its subjects, dives in close, and backpedals seamlessly for the broad shot. As Tim's father had encouraged him to keep tinkering with appliances and his ham radio, Tim in turn supports Paul's passion. In fact, he has put Paul in charge of creating his latest tornado compilation DVD, which he'll sell on his website and at ChaserCon. Since 2004, when Tim put out a two-disc set of his best tornado sightings, called *Driven by Passion*, he has seen a great many storms. It is time, he tells his son, for a new entry, *Driven by Passion II*.

The young man is thrilled to be given the responsibility, and he immediately sets to work wading through untold hours of chase video. He composes ominous musical cues with a synthesizer and cleanly splices together footage shot by Tim, Carl, and eventually himself. Each storm is given a neat opening dateline to anchor the viewer in time and place. Best of all, the episode legend on the inside flap is set against a photograph Paul himself has taken. It is a once-in-a-lifetime shot of a monstrous tornado, folds of condensation and dust spiral up its broad flanks like frayed bandaging. Kathy urges Tim to put Paul's name on the cover. "He needs credit for this," she tells him. Paul is content with his name featured prominently in the credits: "Produced by Paul Samaras."

Driven by Passion II is no Pixar epic; to Paul, the DVD is something more. It is the highlight reel of a chaser's life. An accounting of what his father has done, and how. In the world of storms, Tim is an icon. A friend of the family, Sharon Austin, asks Paul if he ever feels lost in the wide span of his father's shadow. "No," he replies. "I'm just happy to be a part of things."

After immersing himself in Tim's footage, Paul hits the road with TWISTEX more and more often. At first he is relegated to observing from the backseat of the probe truck or a mesonet. But his talent is unmistakable, even if his presence in the group is muted. Paul is a hard guy to get to know, says Lee and Finley's former student Matt Grzych: "He's a shy guy, but an incredible photographer. He took some of the best tornado images I've ever seen."

With the TWISTEX team, Paul is finding not only direction, but a second family. In a video shot by one of the mesonet drivers, Ed Grubb, he and copilot Tony Laubach pull to the side of the road behind Tim's truck, near Campo, Colorado. Hail piles in pristine drifts up to their ankles and smokes in the mid-May air. Tim, Carl, and Grzych compress the ice pellets into dripping balls and hurl them at M3. Hailstones soar back and forth across the ditch in coordinated fusillades. Laubach scores a direct hit against Tim's back. Tim returns fire, and given his form, it becomes clear that his Little League days had in fact been spent studying the clouds. They laugh and holler, and by the end of the fight all are soaked in ice water.

And there, at the edge of the frame, Paul, TWISTEX's documentary eye, peers through the lens of his DSLR, the faintest smirk playing at the corners of his mouth. The image speaks volumes. He's not the guy to join the hurly-burly, but he is one of them nonetheless, taking it all in.

His mother doesn't so much mind the chasing. For all her misgivings, it warms her heart to see how close Tim and Paul have become, united by a common mission. Again she worries when she sees clips from deployments like Tipton, where the margin of error seems so narrow. But then, Kathy knows, she's not the expert. She's never been storm chasing. "I didn't understand how they moved," she says. "I always figured [Tim] knew not to get too close."

Even so, she speaks in no uncertain terms about the ground rules when Paul joins TWISTEX—never risk their child's safety, no matter how worthy the cause. Tim understands, in the marrow of his bones, but she drives her point home with unmistakable clarity: "You have my only son."

CHAPTER SIXTEEN

WARNINGS

BY THE TIME VORTEX2 is ramping up for its debut in 2009, TWISTEX has already gotten a serious head start. The 2008 season was a bountiful one for Tim's team. The Quinter tornado was a bracing experience for Lee and Finley, but the storm also contained some of the warmest temperatures they've ever heard of inside a rear-flank downdraft. Sifting through the mesonet data, they can only describe the quantity of latent heat inside the RFD surge as "astronomical." "Clearly, it's getting air from some other source, and it was dramatic," Lee says. This sample seems to bear out their hunch that the warm downdraft air isn't falling so much as being *driven*. If the tornado's explosive growth following the surge is any indication, Lee and Finley have amassed their first well-recorded case in which this unique downdraft has been associated with the drastic intensification of an already violent storm.

The week following Quinter, Tim celebrated the intercept near Tipton, Kansas, with Paul and Carl. The media probe logged a direct hit, and a nearby turtle's pressure trace seems to have detected a series of secondary vortices following on the heels of the EF1 tornado. The findings derived from HITPR, the media probe, and mesonet data will be published in the American Meteorological Society's prestigious *Monthly Weather Review*.

From the beginning, TWISTEX's objective has been to observe all range of storms—monsters like Quinter, and even relatively minor events like Tipton—so that its members might better understand the features these tornadoes share and, perhaps just as important, the ones they don't. The group is already making progress.

But on the cusp of the 2009 season, the well runs dry. For the first time since the earliest days of the turtle, the National Geographic Society has declined Tim's grant request. A relationship that has lasted for six years ends, for the time being, rather abruptly. "Obviously, it was not an easy decision," says Rebecca Martin, director of the society's Expeditions Council. "But with limited funds we do have to make hard choices."

For the moment, Tim can still count on Bill Gallus as an ardent supporter, though Tim knows that Iowa State's NOAA grant will not field the entire team. He has gathered other revenue streams over the years—small endorsements and merchandising deals—but they are usually made in exchange for free gear, not operational cash. Unless he can round up a new sponsor, the future of TWISTEX suddenly looks uncertain.

As if one calamity weren't enough, the bad news lands at a time when Tim's probe program seems to be faltering as well. The instinctual calculus that guides Tim in the field is lately placing him at odds with his scientific mission. Deploy too early and he misses the tornado. Deploy too late and he doesn't get out in time. His talent has always been for straddling the line between. He takes into account the road network, the tornado's trajectory, and its forward speed. And unless he arrives at something like certainty, he won't step out onto a patch of ground that in minutes—or seconds—will be exposed to winds capable of removing a house from its foundation.

Bill Gallus understands that this means the deployment of certain devices must be prioritized above others. But in recent intercepts, he hasn't been able to glean the data he needs. His task as Tim's brain trust is to gently push him toward the instruments that offer the most scientific value. The trick is to do so without offending him in the process, as Josh Wurman had at ChaserCon.

Like Wurman before him, Gallus wants wind-speed data, not pressure. There's a big question mark surrounding the distribution of wind speeds in the boundary layer, and that's where the most promising research lies right now. The media probe is a better tool for gathering velocity estimates, especially as the wind turns vertical, since it's possible to use the video to deduce the speed of debris. Even more reliable is a *pair* of media probes, which is what Gallus would like to see more often. As subtly as possible, he presses: "Maybe there's some chance you could deploy the media probes first?"

For reasons both romantic and tactical in nature, Tim has insisted on prioritizing the deployment of HITPR. The original probe is his first love, and pressure is his longest-running data set. It brought him acclaim, financial backing, and the respect of scientists he admires. There's a certain stubbornness and pride to his trust in the turtle. More practically, the turtle weighs only fifty pounds and can be deployed in moments. The media probe, on the other hand, weighs a backbreaking one hundred pounds. Tim is now fifty-one, doing most of the heavy lifting himself. At moments like this Tim must empathize with Howie Bluestein's travails during the TOTO years, bogged down not only by the wiliness of tornadoes, but by the weight of his own instrument.

If there's a third parcel of bad news to complete the trio, it's that Josh Wurman might have been right after all. His words from back in Aurora—about the limitations of Tim's turtles, and the precariousness of his funding—must start to echo now. Both warnings are suddenly, aggravatingly prescient.

Tim has spent the last few seasons with his nose to the grindstone. If he's to guide TWISTEX toward major breakthroughs, he needs to look beyond each day's chase to next month, next year. What are they doing right? What are they doing wrong? Scientific leadership requires a whole new state of mind for Tim, and he's been caught off-guard. To be effective again, Tim's probe program needs a new direction. The turtle and the media probe were both revolutionary in their own ways, but a different device is required to push atmospheric science forward long-term.

More immediately, Tim needs a new source of operational cash. With Wurman's VORTEX2 experiment about to kick off in May, TWISTEX can't very well prove itself if it can't afford to chase.

Ironically, it's Wurman himself who opens the door for Tim's next opportunity, perhaps inadvertently. In December 2008, the Discovery Channel reaches out to Tim and invites him to join the cast of *Storm Chasers*, a prime-time meteorological melodrama that tails researchers and severe-weather junkies on the cross-country hunt. With no other untapped benefactors, it's an offer that Tim can't possibly pass up.

His soon-to-be costars, Reed Timmer and Sean Casey, pilot custom-armored vehicles into proximity for the ultimate shot. They shout, jockey, engage in internecine squabbles, and make for divertingly watchable reality television. Considerably less theatrical is the team led by the show's only scientist, Dr. Josh Wurman. He deploys mobile radar trucks and pod probes in pursuit of data, a cool, collected quest not so readily suited to its fleeting slice of a forty-minute show. "The joke about our team was that we're sort of serial killers," Wurman says. "Our pulses don't go up when the tornado happens. The Discovery people were thoroughly horrified."

So horrified, in fact, that the producers need another research group to join the cast and leaven Wurman's undramatic sobriety. To keep TWISTEX afloat, Tim agrees to sign over to Discovery the exclusive right to film his exploits for the show's third season. With Discovery's money, he can now afford to field his team for the entire season. There will even be enough left over to fill out the roster. He brings mesonet driver Ed Grubb on full-time. The fifty-five-year-old pensioner from Thornton, Colorado, a suburb of Denver, possesses a well-combed and enviable head of salt-and-pepper hair, and a spiny mustache above a perpetually mischievous grin. At heart, Grubb is still the young goof-ball who played wide receiver for the Colorado School of Mines. He studied petroleum engineering there, but his career didn't survive the oil-price collapse of the 1980s. After that, he switched gears and hired on with the Adams County School District in Commerce City, Colorado, as head of maintenance and construction. In his retirement, he chases storms and has probably seen more tornadoes than just about anyone else on the team. Like Tim, he may not have earned a degree in meteorology, but he knows how to keep a weather eye on the dangers ahead. "He's a good guy to have out there," Bruce Lee says, praising his situational awareness.

Under the watchful eye of the Discovery Channel, the 2009 season gets under way. The early months offer few useful intercepts, but things pick up in June. On the nineteenth, Tim at last lands the double media-probe deployment Gallus has long begged for, just outside Aurora, Nebraska. But to his profound disappointment, the footage is useless. While the single media probe's 2004 penetration of the F3 in Storm Lake, Iowa, occurred in broad daylight, the Aurora tornado recording takes place at dusk. The light is too weak to accurately track debris. As it stands, they are no closer to getting a pure wind-speed measurement out of the media probes. The one point of light is Lee and Finley's mesonet work. They collect a data set from across the hook echo of the Aurora tornado over nearly its entire life cycle, a rare thing in the history of atmospheric field science.

But in the meantime, VORTEX2 lands its first big data set of the year on a moderate EF2 tornado in Goshen County, Wyoming. The team surrounds the storm with mobile radars, mesonets, weather balloons, and various in situ instruments. None of the pods lands a direct hit, but the sheer number of observing platforms makes this the most thoroughly examined twister in history.

The 2009 season closes without any further successes, either for TWISTEX or VORTEX2. In a calm year, the first in which both teams have been active, the score ends 1–0, VORTEX2 the winner. With just one resounding success, Wurman no doubt would have preferred more. But it's a good start for the program in a modest season, and 2010 promises a second chance. Tim, meanwhile, still has serious problems to solve.

In the interlude between seasons, Tim begins to bend his engineer's mind to the task of envisioning a new direction for his probe program. He is a man pulled in many directions, by family, by a demanding full-time job at Applied Research Associates, and by TWISTEX. After his day job, after dinner with his family, Tim descends into the basement of his Lakewood home—Kathy calls it his "sanctuary"—and begins to conceptualize and construct a brand-new probe. Rare are the moments when his children find him relaxing on the couch.

———

By early May and the start of the 2010 tornado season, Tim emerges from the basement, confident he has remedied most of the design flaws and limitations of both the media probe *and* the turtle. He has done so by making something entirely new: a next-generation probe that combines both his signature devices—and then some—into one package. Tim dubs the instrument TOWER. Like HITPR, TOWER's base is squat and safety-cone orange, but pyramidal rather than conical. There, the similarities end. The remainder of its instrumentation is mounted on a metal mast rising to the height of the average NBA point guard. Rather than sampling wind speed just a few inches above the surface, TOWER's sonic and conventional anemometers obtain measurements at three different elevations, up to six and a half feet off the ground. At the top, a video pod scans a full 360 degrees. The humidity and temperature sensors, instead of cooking inside the steel shell of the turtle, are aspirated by a small fan.

Its final and most theatrical component is inspired by a previous chase in the Texas Panhandle, where Tim watched lightning set an oil well ablaze. Tarry gouts of smoke had become entrained in the storm's updraft, spiraling skyward. Hoping for a similar effect, Tim has affixed a battery of orange smoke canisters to the center of TOWER's stem.

Without a doubt, this is the most advanced in situ probe ever constructed—as far away as an instrument can get from a fleet of Chevy probes.

But with all its improvements, the package tips the scales at an immovable four hundred pounds. One of the media probe's chief liabilities—its weight—has now multiplied. No one will be hoisting this instrument and jogging down the road with it. Tim will have to deploy directly from the back of the truck.

This means that his faithful black GMC has outlived its usefulness. Tim's new probe vehicle—outfitted by a company in Englewood, Colorado, not far from his home—is the ultimate storm-chasing rig: a white, one-ton GMC diesel, sealed from bumper to bumper with an impact-resistant elastomeric Line-X coating. A brush guard shields the front end, to which eight high-intensity driving lights are mounted. To fend off sharp debris, the off-road tires boast an internal stiffening made of Kevlar.

After the Quinter debacle, Tim also incorporates a winch with 8,000 pounds of pulling power, along with a few new deluxe touches. To the hood, as a side project for Boeing, he has installed carbon-composite tiles and piezometer blocks, for measuring the force of hail impacts. Any time these sensors and tiles take a hit, they will trigger a Vision Research Phantom V12 high-speed camera mounted to the dash. This instrument, which costs about as much as a new car, can take high-def footage at a blistering 6,800 frames per second.

Last, and most essential, is the 1,500-pound-capacity hydraulic lift gate at the rear. This he will use to lower TOWER to the ground—as quickly as possible. Winnowing down deployment time is imperative. "What I'm hoping for is around twenty to thirty seconds, which is quite a task given its weight," Tim says. "If we can do that, we can collect more information than any instrument that has ever been placed in the path of a tornado."

Like a NASCAR pit crew, Tim and Carl drill, deploying TOWER repeatedly. They unfasten the black straps crisscrossing the base, shove

TOWER onto the lift gate, lower it, and pull the instrument off the platform with a pair of steel handles.

Yet no matter how many times they practice, the unloading always takes longer than Tim would prefer. TOWER is as unwieldy as TOTO. Certainly, it will allow for no time to deploy a second device, much less a fleet of them. To make matters worse, TOWER isn't technically ready as storm season rolls around; Tim hasn't had time to install a dynamic pressure-reduction port. But the season is upon them, whether Tim and TWISTEX are ready or not.

CHAPTER SEVENTEEN

BOWDLE, SOUTH DAKOTA

O N MAY 10, 2010, TWISTEX misses the year's first major tornado outbreak, losing out on two EF4s in the Oklahoma City area. Four days later, they're in deep West Texas, pursuing a supercell near a town aptly named Notrees. But the road grid in this part of the state couldn't be more ill suited to storm chasing. TOWER never gets anywhere near its target.

To miss a trio of strong tornadoes with the first new probe Tim has developed in seven years is grating. As much as he knows by now that patience—for weeks, for months, for years—is the key to finding the right intercept, that doesn't make the close calls any easier to bear. Tim's lugging around either the future of tornado research or a four-hundred-pound scrap heap in the back of his truck—and he can't know which until it enters the crucible of a plains titan.

By May 21, 2010, Bruce Lee and Kathy Finley have to return to their home state, for what should be a short breather. They stop off in Minneapolis to visit Finley's sister and complete repairs on a mesonet station that a student had accidentally smashed against a motel awning. In the early-morning hours of the twenty-second, they examine the latest weather model runs, as they have every day so far this season. To their shock, the outlook for the day's storm potential has improved tremendously overnight. The models predict a highly unstable atmosphere over South Dakota. Along with preconditions such as wind shear, they look for the presence of moisture, heat, and the tendency for air to rise, which is accounted for by a metric known as convective available potential energy, or CAPE. With CAPE at 1,000 joules per kilogram, storms are likely. At 2,500 joules, storm chasers start to perk up. For May 22, the models project *4,800* joules per kilogram. This should bring chasers streaming from all directions. It is the kind of number heralding days of legend, the likes of Jarrell, Texas, 1997, or Bridge Creek, Oklahoma, 1999.

Lee and Finley jump on the phone at 6:00 a.m. The rest of the team is currently snoozing in a motel in Chadron, Nebraska, more than 600 miles to their southwest. They manage to raise their former student Matt Grzych, and tell him to boot up his laptop to look at the night's model runs, especially the Rapid Update Cycle, or RUC. Grzych pores over the data and agrees: this is the setup TWISTEX has been waiting for. Before long, Carl, too, is enthusing over the RUC model. That each forecaster on the team is zeroing in on the same South Dakota target is telling.

Grzych and Carl start rousting the sleeping crew. Lee and Finley hurriedly finish the repairs on the mesonet and drive some six hours west onto the South Dakota plains, arriving after lunch with plenty of time to spare. The eastern and central portions of the state are a mosaic of tallgrass prairie and shallow wetlands. Skeins of waterfowl lace the sky. The fields are alive with pheasants as Lee and Finley drive past, sporting so many that they fear one will fly into the mesonet and damage its sensitive sensors. The two spend the remainder of the early afternoon birding, until the magic hour arrives and the cumulus towers begin their initial ascent.

By midafternoon, the rest of the team converges on the rendezvous. Chris Karstens, the Iowa State grad student, always knows when he is within range of his fellow researchers. The silent TWISTEX frequency suddenly crackles to life with the voices of friends, trading joking insults and comparing observations.

Tony Laubach follows close behind in M3, along with Ed Grubb. Following the deployment briefing, the TWISTEX crew saddles up, confident in the day's prospects. TOWER is securely strapped to the back of Tim's probe truck. The mesonets are manned by some of the most competent chasers and atmospheric scientists in the field. The clouds are primed to explode. "It's like you're going to war," Karstens says.

Best of all, they learn that VORTEX2 is sitting this storm out. The night before, the field coordinator's vehicle had broken down. "In their defense, they had been going very hard. They didn't get back to their hotel until 4:00 a.m.," Finley says. "Everybody was pretty exhausted."

With its unmanned drones, fleet of mobile radars, and vast mesonet array, VORTEX2 can cover every facet of a storm in a way the much-leaner TWISTEX never will. Since 2009, VORTEX2's armada has seemed omnipresent, smothering every chase. "Everywhere we went," Lee laments, "there were these VORTEX vehicles. And as soon as they arrive, our data isn't worth a lot."

But the project's very enormity also means that any number of moving parts can break down. Wurman's contingent is a cruise ship—slow to get moving, slow to change course. If the TWISTEX crew had thrown in with VORTEX2, they would have missed what is shaping up to be the most promising storm since the project launched the year before.

With their rival off-line in North Platte, Nebraska, there is no one else with whom to share the roads. The storm is TWISTEX's for the taking. Despite the long odds of Tim's gamble—the limitations of staying small and agile, the money problems, the risk of operating without Wurman's mobile Doppler—it looks as if Tim has pulled it off again. If the new probe works.

―――――

Around four that afternoon, west of Bowdle, South Dakota, the mesonets travel north in a line, roughly a kilometer apart, down a narrow and winding dirt road. Consistent spacing between mesonets is paramount for data collection, and they maintain this distance with a GPS program Karstens has developed, called Sidekick, which displays the location of each vehicle on a map.

Lee can already sense that the day is about to get interesting. The mesocyclone to his west is pendant above the low rise of northern plain like a glacier, jagged and massive and all but kissing the earth. As the fields wash past, and the cars and radio waves fill with excited chatter, Grubb thinks he may have spotted telltale dust swirling along the ground. In M1, Lee identifies the storm's clear slot, connoting the passage of the rear-flank downdraft. The updraft is organizing, shaped and smoothed by laminar winds. The show is about to begin.

"Okay, mesonet," Lee says, "the way we're gonna line up is [M1] will take the lead position. We'll try to stay roughly half a kilometer to the south of the tornado path. Tony, you're a kilometer behind us. M2, you're a kilometer behind Tony."

Arrayed along the flank of the storm, Lee orders the convoy to a momentary halt as they approach the mesocyclone. Any farther and M1 risks straying into no-man's-land, the perilous pocket inside which the tornado could come down quite literally on their heads.

Without warning, a gust wails out of the west and rocks the mesonets, peppering the vehicles with rain and dust like bird shot. They've just penetrated their first RFD of the day. M1's mesonet station reports a forty-six-mile-per-hour gust. The wall cloud—the low-slung feature at the bottom of the mesocyclone, from which the tornado will soon emerge—swings past M3, turbulent and ragged, like dirty cotton. On closer inspection, it appears to contain a possible funnel, or at least the earliest stages of one.

A palpitating lobe of condensation hovers just above M1, an indication that the mesocyclone is strengthening and accelerating. It's preparing to focus the broadly distributed power of its slow turning into the tornado's vicious knife edge. The storm bears away east-northeast, and the convoy begins to move north with it, proceeding deliberately, at no more than twenty miles per hour. As Lee and Finley scan the wall cloud, they notice again the lowering to their four o'clock. Tornadogenesis has a way of feeling both gradual and surprisingly sudden. The feathering lobe of condensation narrows, sharpens, and smooths until at last the tip is honed to a needlepoint. It swoons breathtakingly close to earth, then decays in filigreed shreds before retreating back into the wall cloud. Striated bands of stratus flood the swelling mass of the mesocyclone. The storm is feeding. "Wow," Lee gasps. "This is *something*! We've got winds holding at forty [knots] out of the westnorthwest."

At a rate almost imperceptible to the eye, the funnel returns, building and lengthening, stretching inexorably toward the surface. "The thing is just cranking," Finley says.

Now several funnels knife through the lowering. Is this the birth of the multiple-vortex? The radio waves are chaotic with real-time speculation. But within moments the sky has answered: the multiple vortices coalesce into a wedge. A welter of dust erupts from the fields, and the first reachable tornado of the season has officially touched down. Within seconds it becomes apparent that TWISTEX has a high-end twister on its hands. No need to cycle, to mature, to build angular momentum. This thing was born a classic plains beast. A voice over the radio crows, "Tornado on the ground, folks!"

A cheer goes up over the TWISTEX frequency.

———

North of the mesonets, on the other side of the tornado, are Tim, Carl, and Grzych—packed into Tim's new rig with a Discovery cameraman. They hurry east along Highway 12 toward Bowdle. The vortex drills down onto the horizon, tethering earth and sky. "This thing is moving northeast," Tim says. "It's coming toward us."

"Wow," Carl exclaims, "it's right there!"

The tornado churns along the wide South Dakota prairie, closing in on a slice of highway that'll be their best shot at a deployment. Even with the new probe truck and its big diesel, Tim is struggling to head it off. "The thing's gonna translate across the road," he says.

"Are we going to deploy the probe here?" Grzych asks.

"I'm not sure. I can't get a handle on the direction."

But this is it, now or never.

Carl brings the truck to a stop in the breakdown lane and leaves the engine running as they step out, clothes snapping in the wind. Tim kneels beside the truck, strokes his chin, and looks for the sign. The cloud motion overhead is vigorous, boiling. The tornado has momentarily roped out, but there is little doubt that it will return. The question is when.

Tim peers up at the wall cloud, hovering like a mother ship. He sees low-level moisture drift beneath the updraft like tatters of low fog. This is good. It means the wall cloud is cycling, gathering its strength for the next pulse. The storm could produce another tornado any second now.

The internal clock that guides Tim's deployment timing is ticking down. The circulation is closing in. He turns to Carl and Grzych. "Let's do it."

They remove the tie-downs, lower the lift gate, and strain to pull TOWER onto the asphalt. Tim yanks the smoke-canister pins and orange braids belch forth and stream across the field toward the storm. He raises the lift gate and stands watching for a moment until he realizes how quickly the developing circulation is narrowing the distance.

It looks like the storm is falling from the sky. The wedge descends and vortices start daggering from its underside.

Leaves and wheat stalks bombard them. The men scuttle around the truck and pile in. The cab becomes a wind tunnel as inflow blasts through the open doors. "Get in! Get in! Get in!" they shout at the Discovery cameraman, his lens still trained on the tornado.

As the cab's detritus takes flight, they muscle the doors shut. Carl pushes the accelerator to the floor. The diesel engine roars.

Grzych peers through the rear window at the receding TOWER and the strengthening vortex.

"It's going in," he announces. *"It's going in!"*

Carl raises his fist and the cab rings with joyful hollering.

"All right! All right!" Tim says. "Carl, pull over! Pull over! We gotta record it!"

A short distance up the road, Carl again swings into the breakdown lane and throws the truck into park. The men spill from the GMC. They can see the inflow coursing low over the hay fields—a deadly river of radial wind known to chasers as the ghost train. The funnel isn't quite fully condensed, more like a wedge in gestation, trailing a bolus of swirling condensation over TOWER. In its current form, this may be no Manchester beast, but at least this field test has provided proof of concept: even from here, Tim can see that TOWER stands stubbornly rooted to the asphalt.

Tim's voice rises above the deafening wind almost to the breaking point, his composure utterly lost now. *"It's right in there! It's right in there!"*

He jabs his fists at the air and throws his arms roughly around his comrades in the bar ditch. Over the years, Tim's fortunes have risen and fallen. The creation of TOWER, the most technologically sophisticated in situ probe ever devised, was the jolt his research program needed. Now it's proven worth it. His invention has shown it can stand strong in the guts of a growing tornado. Tim has a reason to smile and laugh again, like a storm chaser in the thrall of his first intercept.

———

Off to the west, the mesonets steer toward TOWER's position, dataloggers humming. Driving point, M1 has the tornado in its sights. Sitting at the dead center of the highway ahead, Tim's fledgling twister has matured and looms ever larger, its color the pale hue of bone. They're in the presence of the beast now, an enormous incisor sunk into the horizon. Somehow, Lee sounds like he's directing traffic. "Let's keep spread out so we can get a good gradient data set off to the west within this RFD until it crosses the road," he says over the radio. "Then we'll scoot in just south, and underneath it, and back up."

As they move in closer, the span of M1's windshield is almost entirely filled with a vortex whose manifestation from one instant to the next is utterly novel, a transfixing, almost hypnotic sight. It is a reminder that a tornado isn't an object in the same way we think of a stone or a handful of earth. It isn't a thing, it's a process, a wholesale redistribution of pressure and air. The effects of this process, however, are all too tangible. Off to the left, M1 is the first to notice leaning power poles, a defoliated line of trees, a house with an exposed span of attic lumber, and pastures strewn with debris of every size and shape.

Directly across the highway, TOWER smokes defiantly; its anemometers wheel, and the orange canisters suffuse the mesonets with the scent of their acrid orange plume. Another cheer goes up over the radio. Laubach screams jubilantly, almost involuntarily: "Mother-*fucker*!" Tim has done it again.

As they enter the outskirts of Bowdle, the wail of the town's siren presses in through the windows. The storm is continuing to grow; they're on the trail of a giant. Its broad shadow sweeps across the plains to the north of town like something out of a bad dream. At Bowdle's eastern edge, M1 turns north up State Highway 47 for the intercept.

When they gain sight of it again to the mesonets' northwest, the funnel is as wide as any Lee and Finley have ever seen. Conservatively, the tornado may be at least three-quarters of a mile across. At M1's ten o'clock, Lee suspects the thing is tracking a hair north of east and will cross the road ahead. Strung along the highway, TWISTEX closes the distance.

"This thing is absolutely gigantic," Lee says. "We're gonna be careful on this one. This is a high-end deal."

The M1 camera is centered on the skinny strip of bright western horizon, where the sun's light ends and miles of frothing, sea-foam clouds carousel cyclonically, from north to south. But below the mesocyclone's low rim, rain and wrapping condensation hide something even darker. Often enough, the rain briefly clears and they see it, hundreds of yards across, tapered like a spearhead whose tip is driven into the earth. Here, the catastrophic EF4 winds sweep away the feathering edges of the meso and file the vortex smooth. Lee pulls M1 to the side of the highway and instructs M2 and M3 to do likewise.

"It's almost stationary," he observes. He assumes the tornado will cross the road ahead at some point, and Lee has no intention of being anywhere near the thing when it does. The mesonets are now in an ideal position to collect data. He resolves to press no farther. With the sun now entirely occluded by the wedge, it's as if they're witnessing a solar eclipse, the funnel nearly black and its gilt edges flaring gold.

At about this moment, the probe truck passes Lee. They've retrieved TOWER and are angling for a second intercept. Inside the truck, Tim is clearly unnerved. The wedge menaces the highway, as though daring them to cross its path, to step into the threshold of a door that is about to slam shut. Manchester may have been all white knuckles, but to attempt to deploy now would be suicide. "I dunno, guys, this is getting too fucking close," he says. Then, to Carl, who is driving: "There are people everywhere, man. This is too damn dangerous. *Stop!* This thing may translate to the east now. We're just going to have to wait this one out."

Carl pulls off onto the breakdown lane a quarter mile or so ahead of the mesonets—and about 400 yards from the tornado. Without warning, a massive RFD surge rocks the probe truck and each of the mesonets. Rain curtains sweep over the fields and obscure the tornado.

"M2 is picking up a good RFD from the west."

"M1 is, too. Hopefully it doesn't get so strong that it knocks these power lines down."

"The tornado is becoming rain wrapped from M2's standpoint."

"Roger. We've got large debris coming up on the east side of it."

After only a minute, the view clears out again. In M1, Lee and Finley spot the tornado's trailing edge. A single subvortex emerges against the backlighting sun and separates from the mother circulation for an instant. Then it rotates back around the south face, out of sight. Even the hardened researchers are reduced to monosyllabic expressions of awe.

"Wow," Finley says, her voice hushed.

They can see now that a second RFD surge has touched off a transformation. What first appeared as a ragged wedge now more closely resembles an enormous stovepipe, as though a pair of great hands has molded and squeezed the funnel into something sleeker, narrower, and infinitely more violent. Its vortical profile rises a precipitous ninety degrees from the earth below like a superstructure's mammoth colonnade, a true destroyer of cities.

Tim, Carl, and Grzych kneel in the sheltering lee of the GMC. From this distance, they can just make out a row of 160-foot power transmission towers. They must weigh many tons apiece, but next to this thing, taller than any mountain, the towers seem brittle and tiny. As the edge of the vortex moves to overtake the first, the metal spire emits a brilliant blue pulse of light and folds down onto itself. The shrieking tangle of steel beams shears from its concrete footings. One after the next, the towers fall. From its point of anchorage, one of them plows the prairie for some four hundred yards before the wind finally releases it.

The wall moves onto the highway, as Tim had predicted. Then the rain curtain swings back around and shields the tornado from TWISTEX's view. Laubach suggests finding another road to pick up the chase, but Lee urges him to hang on just a little longer. He wants the mesonet sensors to savor the entire evolution of this transformative RFD. There seems to have been a direct correlation between the surge and the tornado's most exceptional intensity. But because Tim had been unable to deploy on the tornado in its maturity, they lack the data to prove that the surge amplified wind velocity within the vortex. Lee has only the feeling in his gut.

Even more striking is the final RFD surge, currently ongoing, which the mesonets are picking up even as the storm moves away. In contrast to the warmth of the preceding pulse, this one is markedly chillier. The change in temperature must mean it originates from some other place in the storm, where a common downdraft's evaporative cooling plays a more prominent role. The divergence both in temperature and strength between this and the previous surge could not be any starker. Yet both originated from within the rear-flank downdraft. How could the same downdraft behave so differently within the same storm? This is what Lee and Finley seek to figure out.

In fact, TWISTEX may now be in possession of something priceless. In the long history of atmospheric research, only two coordinated mesonet data sets have ever been collected from top-percentile tornadoes, and one of them is theirs from Manchester. This is now the third in existence, and it documents more than just one surge. They have slices through a handful of RFDs, each with incredibly disparate effects on the storm. The first accompanied tornadogenesis. The second triggered intensification. The third arrived shortly before the tornado's most extreme damage was documented. And the fourth set in motion its death spiral and coincided with the tornado's turn to the north. Clearly, the surge is some kind of signal, and potentially even the mechanism behind each stage of vortex evolution. Lee and Finley have reams of data to sift through, and a scientific paper to write. In the years ahead, they'll have tantalizing new questions to answer: Is the surge a necessary condition for strong tornadoes? Can it be traced to some other environmental condition that might lengthen warning times?

Bowdle is a heady moment for TWISTEX.

Grzych and Carl's appraisal of the day is less circumspect. The two are positively giddy. "VORTEX2 is freaking *shitting* their pants right now," Grzych says.

"They lost *big*-time," says Carl.

———

The following day, Tim and Carl return to the location where the Bowdle tornado reached peak intensity. They drive down Highway 12 and note the power transmission towers that lie in heaps of line and galvanized-steel lattice. They stop along the road, near what had once been an attractive little pocket of cottonwood trees. Corrugated aluminum is wrapped around the denuded trunks like shoddy armor. They approach what appears to be a sedan, eerily similar to the white Chevy Cobalts used by Lee and Finley's mesonet. The front and rear ends have either been torn away or shoved into the cabin. An impact with some large object has caved in half of the roof.

"What we're seeing here is a vehicle that did not originate in this grove of trees," Carl observes. Indeed, the car had been parked at a home that once stood some seventy-five to a hundred yards away from here.

"It was at some point a missile," Carl says.

"Yeah, this is certainly a good example of why you're not safe in a car," Tim says. "People who try to outrun tornadoes in a car like that? It's not pretty."

CHAPTER EIGHTEEN

A DEAD END,
A NEW CHANCE

I N THEIR TWO head-to-head seasons with VORTEX2, Tim and his team have proven themselves beyond a doubt. While the probes fell short in 2009, the program surged back in 2010. At the same time, Wurman's armada wasn't able to close on any of the season's major tornadoes. For TWISTEX's more competitive members, the underdog crew might've even snuck out as the winner over the last two seasons. Next to their transmission-tower-eating Bowdle wedge, VORTEX2's only complete intercept—the Goshen County, Wyoming, twister in 2009— looks like a dust devil. TWISTEX's sheer agility allowed it to thrive even next to NSF muscle and the finest mobile tech money could buy.

None of their chases, though, would have been possible without *Storm Chasers*. In more ways than one, the Discovery Channel show has been a blessing. It kept the mission on the road when funding had all but vanished. And it brought a handful of more worldly benefits as well. For Tim, as Kathy refuses to let him forget, the show has raised his profile from relative unknown to minor celebrity. He had always been a hit at weather conferences, but once *Storm Chasers* arrived on basic cable, Tim started finding himself approached for autographs in airports. Kathy teases him about his graying hair and asks with mock sincerity whether he should color it before presenting himself to his multitudinous fan base. Tim makes a show of grumbling and says, "If I got gray hair, I got gray hair."

The influx of funding also helps Tim and Kathy move away from their quiet Lakewood street—now too often congested with mesonets and the Discovery Channel support vehicles—and into a palatial brick compound with thirty-five acres in the foothills east of Denver, near Bennett. Tim found the new place by trawling foreclosure auctions, and the move is a long time coming. For years, he had lamented to Kathy that they'd outgrown Lakewood, that there was no room for a real shop, not to mention that the city had refused to grant him approval to lengthen his hundred-foot ham radio tower.

It's been a constant dream of Tim's to own acreage on the plains, where he can erect an antenna farm and chase storms to the east. Even so, the house is unlike anything a middle-class guy raised in a small bungalow would have dared to envision for himself. The couple's new master suite has a second-floor balcony with stunning views of the Front Range in the west, and the high plains to the east. The entryway and the living room are bathed in the natural light of huge picture windows opening onto a two-story atrium. Though its extravagance appears at odds with Tim's unshowy practicality, the house also includes several geeky amenities that appeal to him more than any luxurious frills: The basement is vast enough for a hoarder's profusion of tools and gadgetry. And the expansive four-car garage will become the perfect setting for a state-of-the-art shop, replete with band saws, lathes, drill presses, a Miller arc welder, and a custom vacuum system that can keep the floors clean and fireproof by suctioning and separating hot metal slag and sawdust.

The initial transition from cozy bungalow to echoing mansion is a little jarring. *This isn't us,* Kathy thinks at first. But now that they're settling in, she and Tim are both getting used to it. From their bedroom balcony, the Milky Way lights up the nighttime sky, and the sunrise unfurls a hundred miles of plains before them. It is no exaggeration when Tim looks out and tells her, "I can see thunderstorms in Kansas."

Yet, as many doors as the Discovery Channel has opened, there are catches to contractually obligating oneself to reality television. They reveal themselves in due course.

The most obvious and expected are the cameras. For at least two months of the year, Tim now lives under a microscope, with a lens never more than a few feet away. From dawn to dusk, the cameras hover near. Curiously, the camera operators and producers have chosen to hew closely to Tim and Carl, to the exclusion of the rest of the team. Laubach gets a line of dialogue here and there. And given his excitability, Matt Grzych is installed in the probe truck by the producers. For this, Grzych and Laubach are paid talent fees. The expenses for the rest of the team—gas, lodging, food, equipment repairs—come out of Tim's pocket. As far as the show runners are concerned, Bruce Lee and Cathy Finley's mesonet fleet—fully half of TWISTEX—does not exist. "They don't want anything that looks like Tim isn't the only leader of this project," Lee says. "They have their own story line. We basically caused them a whole lot of editing."

Otherwise, the first two seasons of shooting have transpired much as the TWISTEX team had anticipated. *Storm Chasers* adheres to a standard Discovery Channel reality formula. The score is all electric-guitar riffs and cymbal crashes. The sound engineers occasionally insert some stock wildcat roaring and growling to augment the moaning of the wind, as if a hundred-mile-per-hour gale weren't fearsome enough. The interteam tension is usually manufactured, a ginned-up horse race to spot the most and biggest tornadoes. Even Tim seems unable to resist the temptation to play to the cameras, at one point intoning ominously, "This could spell *disaster!*"

Aside from a few harmless dramatics, though, he refuses to lob snide sound bites at his so-called competitors or to indulge in intrateam sniping. Tim instructs the Iowa State students to behave professionally. "They're going to create stories that do not exist because it's a TV show," he says, "and it has to be about catfighting. You guys will not talk about stuff if we're having disagreements."

For Carl, *Storm Chasers* is his biggest role since his fleeting appearance at the end of *Against the Law*. He gives himself over to the process with gusto, even going so far as to enroll in voice-acting lessons. The premiere is a major event for him and his father, Bob Young, and they follow the show religiously.

For much of the crew, the hovering cameras, the dumbed-down reality format, and the wholesale excision of the mesonets are little more than annoyances. They've borne them because TWISTEX can't survive without Discovery. But gradually, minor grievances give way to legitimate problems. For starters, funding seems to arrive at the last minute every season, forcing Lee, Finley, and Tim to calibrate gear in the field or risk losing deployment opportunities. "Motel parking lots aren't the ideal laboratories for installing new equipment," Finley says.

Without fail, there's always some component that isn't ready. Tim's TOWER deployment on the Bowdle, South Dakota, tornado last year, for example, could have yielded much more about the thermodynamic characteristics of the developing funnel. Unfortunately, TOWER couldn't accurately record an essential metric—absolute pressure. In the rush to prepare for the season, Tim didn't have time to install a dynamic pressure-reduction port. This would have allowed the barometer to measure something other than the pressure exerted on the sensor by the wind. Without the port, getting absolute pressure is akin to taking the temperature of a kitchen while holding a thermometer directly above the heated stovetop. "Every year we spent half of the season getting ready for the season," Laubach says. "Fixing shit, working on shit, missing days, missing tornadoes."

By 2011, TWISTEX's third season on *Storm Chasers*, tardy funding isn't their only gripe. The producers' contrivances have become more and more brazen. "If they weren't happy with a scene," Grubb says, "they'd make us redo it. We were at their mercy." The producers even try to feed Tim lines, according to Kathy and others. "He didn't like that at all," she says.

Worse are the expository after-action interviews, during which a member of the Discovery crew would pull Carl, Grzych, or Laubach aside and attempt to coax out criticisms of fellow teammates. Harmony, after all, doesn't make for a good story line, while conflict breeds narrative. On national television, Carl declares more than once that Tim's overcaution is costing them probe deployments.

But Carl, too, feels the team's frustrations. In private conversation, he confides to a girlfriend, Melissa Daniels, that he feels manipulated, like a puppet being pulled about by its strings. One evening, they attend a concert in Carmel Valley, California, where the composer Philip Glass plays a piece from *The Truman Show,* a film about an insurance agent living in a small-town idyll, only to discover that he is the unwitting star of the world's most popular reality television show. Carl tells Daniels he can relate.

More than sowing discord within the ranks of TWISTEX, the show runners have succeeded in drawing the team's ire upon themselves. When Finley and Lee approach Tim's truck to consult with him on a coming deployment, a producer attempts to hustle them away, grousing that they're ruining the shot. Tim is apoplectic. "This is a science mission first," he lectures. "Don't you be shooing our people away."

Laubach sees the pressure wearing on him. "I'd never seen Tim really, truly annoyed before. Like, *I'm fucking over this thing.*" The show and the mission to which Tim has dedicated his life feel increasingly at odds.

The event that epitomizes the growing gulf between research and reality television comes when the models indicate a potential outbreak in the southeastern United States in late April 2011. Despite Dixie Alley's abundance of exceptionally powerful tornadoes, the Southeast is not a region in which TWISTEX customarily operates. The storms often course across the knotty landscape at up to highway speeds, turbocharged by a 150-mile-per-hour jet stream. If the screaming pace of the storms doesn't make it difficult enough to keep up, the hills and trees render southeastern tornadoes nearly impossible to track. In such terrain, Tim would normally turn to radar for guidance. But wireless coverage is spotty at best; the backwoods can be a data black hole. And unlike the reliably gridded road networks of the plains, those in the Southeast lack any dependable coherence.

"There's no point," Lee says about chasing in the Southeast. "It's just hills and forest. It's bloody dangerous." Even if they do manage to deploy on a tornado, the eddied turbulence shed by trees and hills would muddy up the data. By now, though, it has become clear that data collection is not the show's primary concern. The producers call the shots, and today they want TWISTEX to go East.

Tim stifles whatever misgivings he has about venturing into Dixie Alley and drives east with Carl and Grzych. Grubb, Laubach, and Paul Samaras decide to come along for the ride, following behind in one of the mesonets. They're all weighing the risks against the potentially historic significance of the outbreak. Lee and Finley decide to sit this one out.

————

On April 27, near Aliceville, Alabama, no one is getting any signal, and the last radar update is nearly thirty minutes stale. Even Tony Laubach, a mesonet driver renowned for his tolerance for hairy intercepts, is uncomfortable. "This is getting stupid," he says.

The experience of chasing in the densely wooded Southeast feels alien to Tim. On the plains, there is little to obstruct his view apart from the occasional line of cottonwoods, or a low rise that will pass in a moment. In Kansas, there are places where, if you squint hard enough, it feels like you might be able to see clear to the rim of the world. But in Dixie, it's like driving through a narrow hallway. The matchstick pines are close and impenetrable. Descending into the deep hollows, a chaser feels caged. Based on the last radar scan, they should be getting close to the track of a storm that's been repeatedly producing tornadoes. But they have no way of knowing whether that track has deviated in the minutes since they lost cell service.

Suddenly, they spot a tornado lifting and reforming over the canopy. They can barely see the rest of it through the deep woods, but it must be coming back soon. Thirty seconds later, the vortex thrashes through the trees and spills into full view. It's just 100 yards ahead of them, bursting forth like a thick plume of coal ash to the face.

They catch the thing on camera as it sweeps across the two-lane country road and plunges just as quickly back into the pines. No one argues when Grzych notes that only a few seconds of forward progress separated the probe truck from eternity. The gasp-inducing close call will turn out to be one of the most compelling moments of the season for *Storm Chasers*.

Later that day, near Pleasant Ridge, just west of Tuscaloosa, a violent wedge passes a mile and a half up the road, effectively ending their chase. Trees with trunks a foot wide, tangled with matrices of downed power lines, block their route. The odor of broken pine and hardwood is overpowering.

By the time they find the interstate, Birmingham and Tuscaloosa are disaster zones. Some sixty-five have been killed. The highways leading into and out of the cities are parking lots. Given the tornado's path, Grubb believes they saw the monster responsible a short while before it hit Tuscaloosa.

Over four days, some two hundred tornadoes kill 321 people in five states. Fifteen are extremely violent, rated EF4 to EF5. The outbreak is one of the deadliest in modern history, and TWISTEX is lucky to make it out safely. They're wrung out and frustrated, both by the awful terrain they'd been sent into, and by their impotence in the face of such destruction—not a single probe landed, nor did they ever have a look at a viable intercept.

———

By the September 25 premiere of TWISTEX's third season on the show, the series' popularity—and its quality—have declined. As the first episode airs, Tim suspects *Storm Chasers*' run has come to an end, and the ratings don't improve as the season wears on.

The following January, Discovery announces the show's cancellation. Though it means TWISTEX is once again without a sponsor, Tim greets the news with something like relief. The cameras have worn him down. The gray in his hair, a light dusting at his temples and sideburns in 2009, has since marched past his ears and down the nape of his neck. "I saw Tim age," Kathy says, over the course of those years.

"If you knew him long enough," Lee says, "you knew where his high-energy level normally was, and you could tell it wasn't there."

In early 2012, Tim phones Lee and Finley with the proposal they are all expecting: "How would you guys feel if we took this year off from official operations?" he asks. In truth, there really is no other option; everyone needs a break, and the money isn't there. If funding hasn't materialized by 2013, he suggests they start beating the bushes.

For Tim and his colleagues, this is a time of transition. At Iowa State, Gallus's money has run out as well; 2011 marked the final year he would send students out into the field with TWISTEX. Even Tim's work with Larry Brown runs up against similar budgetary constraints. In 2009, they had moved from Applied Research Associates to another contractor, National Technical Systems. But with Congress at an impasse and the federal budget funded on a continuing resolution, the spigot has been shut off. "So here we are—screwed," Brown says. "We got all these great ideas and couldn't get the money." The hotshot team of explosives experts that has worked side by side since Tim was twenty years old has been disbanded.

Through it all, Tim insists that TWISTEX will return; this is merely an unplanned hiatus, a breather. They've accomplished far too much to quit now. In Quinter and in Bowdle, they came away with some of the most compelling data sets ever gathered from within a rear-flank downdraft. Their coup at Bowdle outshone the whole of the multimillion-dollar VORTEX2 expedition. Lee and Finley have only just begun to tease apart the enormously heterogeneous surges responsible for transforming its vortex. If they can sample a few others just like it, TWISTEX will have cornered the market on data sets within what could be the essential mechanism of the storm. What's more, the team now has at its disposal the most advanced in situ probe in the field. TOWER had proved itself near Bowdle. It may well be the tool they need to capture the data sets Tim and Gallus have been hunting. There will be more and bigger storms yet to come—more Manchesters—for Tim's fully completed probe. TWISTEX's best days, they all hope, might still lie ahead.

But as Tim's team stands down, Dr. Josh Wurman will take to the road on another expedition. VORTEX2 may have stumbled in its final year, but Wurman has landed the necessary grants to extend its mission into a new project, dubbed ROTATE 2012. To add salt to the wound, ROTATE's core mission is to illuminate the near-surface environment in which Tim has specialized for years. The pièce de résistance of the effort is Wurman's fleet of pods, now numbering twenty-two in situ probes, each equipped to take video while also logging pressure and wind speed. While Tim has been forging his new Lamborghini in TOWER, Wurman has built his pack of Chevrolets—which he'll be deploying this year while Tim sits on the sidelines.

Tim chose the gravel road during his meeting with Wurman at ChaserCon all those years ago. He's scraped by and strung along, making magic out of nothing. But now all of Wurman's warnings have held true. The longer Tim pushes forward, the clearer the destination becomes. After all this time, his route is looking like a dead end.

There's no time to wallow, though. Tim must attend to the present. Out of work and now semiretired, he approaches a company in Tupelo, Mississippi, that had built a small mobile radar for Reed Timmer during *Storm Chasers*. Hyperion Technology CEO Geoff Carter is already a big fan of Tim's. An employee who'd been in the field during Discovery filming had returned to Tupelo gushing about Tim: "We gotta hire him; he's a genius." Carter offered him a job at the time, but Tim had too many obligations then.

The offer, Carter tells him now, is still on the table. But before Tim accepts, he has a couple of unorthodox conditions: First, he won't relocate to Tupelo and uproot his family. Second, he wants a guaranteed leave of absence in the early spring and summer to field TWISTEX. Whether the team takes to the plains again may be an ambiguous prospect, but Tim isn't prepared to let it go. Storm season is sacrosanct and always will be.

Carter doesn't hesitate to agree to the terms.

Hyperion's ongoing research may have nothing to do with tornadoes, but one project at least falls within Tim's sweet spot at the intersection of tech and severe weather. The company is developing an instrument package for NOAA that will measure hurricane wind speed and alert coastal communities to storm surge. Carter's engineers have been racking their brains to come up with a sensor that is not only capable of reliably detecting the rising waters, but tough enough to take a sustained pounding. Tim's off-the-cuff idea is practically antediluvian in its simplicity. Place a tube in the water and cap off the top with a pressure transducer. If the water rises, so will the pressure. The ease with which he has solved the device's technical challenge is stunning. Carter even lets Tim deploy the package, dubbed eyeSPY, on the Mississippi coast, near Waveland. Hurricane Isaac is already darkening the Atlantic sky just off to the east.

Next up for Hyperion, Tim takes on the construction of a new chase vehicle, one unlike any the world has ever seen. In 2010, he'd been contacted by an atmospheric scientist named Walter Lyons, who made an offer an intensely curious man such as Tim couldn't possibly refuse. "DARPA said, 'Send us your ideas; we want high-speed photography of lightning and sprites,'" recalls Lyons. "I needed a partner. Tim was the logical person." During a 2006 side project for NatGeo, Tim had single-handedly documented aspects of lightning behavior that physicists had only theorized about, using the same high-speed cameras he'd trained on explosive shock waves for decades. When Lyons reached out in 2010, the pair spent the off-season shooting lightning, but found that the risk of stepping outside a vehicle amid deadly strikes was too dangerous. To succeed, they concluded, they'd need to find a way to stay inside. Now, Lyons and Tim have a chance to make good on that idea—with a recent influx of funding for lightning research and a blueprint for a customized lightning-chase vehicle.

Like the sprawling VORTEX2, the project that Lyons wants Tim to join is a moon-shot effort underwritten by the federal government to unravel the fundamental mysteries of lightning. The project isn't interested solely in the familiar bolt, but in observing a menagerie of dimly understood electromagnetic phenomena called transient luminous events, or TLEs. During some thunderstorms, electrical discharges splash luminous plasma across the upper atmosphere like paint in one of Jackson Pollock's expressionist works. They can span dozens of miles but last for only a fraction of a second. Researchers want to know why only some lightning produces TLEs. Of equal interest are the recurrent upward strokes emanating from wind turbines, and the expensive repairs they often necessitate.

The effort, dubbed the Physical Origins of Coupling to the Upper Atmosphere from Lightning, or PhOCAL, will involve a massive Lightning Mapping Array. This network of antennas, electric-field sensors, and continuously recording video cameras are spread across sections of central Kansas and west-central Oklahoma, usually near wind farms. SpriteNet cameras have been installed in locations throughout the plains, including one on Tim's bedroom balcony.

Tim's contribution to the effort is what he dubs the Lightning Intercept Vehicle, or LIV. He'll spend much of 2012 and 2013 building it with Lyons and Hyperion. Designed specifically to capture nature's most transitory occurrence, LIV looks like your average white box truck. Upon closer inspection, however, its innards brim with the most advanced high-speed cameras money can buy. Hyperion donated the turret at the top of the truck, from which two Phantom cameras rotate 360 degrees. "You could buy a nice Bentley for what you'd pay for one of those Phantoms," Lyons says. Mounted to the sides, front, and back, four closed-circuit, black-and-white Watec cameras are angled at the upper-atmosphere to capture sprites. A computer and a bank of screens in the back of the truck run the various cameras and a satellite weather system for lightning detection.

But the real crowd-pleaser is the modified Beckman & Whitley Model 192. The Kahuna, as Tim refers to the instrument, is a nearly one-ton Cold War relic formerly used by the military to photograph nuclear explosions. It's the videographic equivalent of a Gatling gun: eighty-two cameras spinning around a drum at up to 4,500 rotations per minute. The Kahuna can only run for thirty seconds or so before it overheats and tears itself apart. "When you approach eight hundred thousand, nine hundred thousand frames per second," Tim says, "it screams like a girl."

Getting the contraption into the back of the LIV requires nothing less than a forklift. Over the years, Tim has developed something of an emotional attachment to the Kahuna. At ARA, he retrofitted its film cameras to shoot digital images, an enormously complex undertaking. He now believes the hulking camera will reveal lightning behavior scientists haven't even conceived of. "The objective there is to catch the attachment process, where the step leader touches the ground," Tim says. "I want to see that formation process. I want to see the formation of the return stroke coming up off the ground."

If there is anything that might be able to compete with Tim's fascination with the vortex, it's lightning.

He is already deeply engaged, obsessing over the LIV through the beginning of 2012. As the first severe weather of the year approaches, though, National Geographic comes through with an offer of just enough money to fund Tim's tornado chasing for the season. It can only cover him, not TWISTEX; in exchange Tim will post "webisodes" of his best footage from the road. Tim can't say no to an opportunity to chase.

Soon, he and Paul are on the trail of a powerful EF4, near Salina, Kansas, in mid-April. The wedge roars over the table-flat fields at a crawl, allowing them to approach within a couple hundred yards. But that one chase will have to hold Tim over until next year.

The ongoing drought is cutting the storms off from one of their main fuel sources: moisture. No other significant tornadoes touch down on the plains for the rest of the season. Even Josh Wurman, with all his funding, struggles to find any worthwhile storms. It may be just as well that TWISTEX is taking the year off.

Through the rest of 2012, Tim pours himself into the LIV. It should be ready in time for the 2013 season. He may wonder, as he works, whether he could ease over into a parallel pursuit from here on out: lightning over tornadoes. This could well be the next unexplored frontier. He and Lyons have two years' worth of DARPA funding, and they're already planning a series of lightning-related research projects. Could this be his next Manchester, even?

After nearly twenty-five years of chasing and too many close calls to count, Tim has an off-ramp if he wants to take it. He's fifty-five years old and can devote his time to safer projects—he could embrace the landmark lightning study, lay off heavy tornado hunting, and spend a lot more time with his wife, kids, and grandkids. He has taken a step back; he can turn away entirely if he chooses to.

But the vision of a Kansan EF4, with Paul at his side, doesn't fade so easily.

As much as Tim geeks out over the Kahuna and the LIV, it's not the same. It's not tornadoes. Even as Tim has assembled new projects to support himself, he's still looking for a way to revive TWISTEX.

As the 2013 season approaches, he thinks he might have found it: National Geographic says it is prepared to underwrite Tim's tornado research in a bigger way again. Unfortunately, when the grant is finally funded, he only gets about half of what he'd asked for. The money is enough to keep Tim, Carl, Paul, and maybe another friend in the field. But it won't cover a full complement of TWISTEX mesonets. He writes to Lee and Finley to break the bad news: they're on their own once again this season.

Tim has enough for a small mission, at least. And he's intent on making the most of it. It will be like the old days—back when he chased with a minivan, a few turtles, and a companion to hold the camera.

———

Come early April, the familiar rush is back. Chasers are a helplessly optimistic bunch; the beginning of every storm season holds the crackle of promise. Any afternoon, Tim may witness something incredible, something that casts a shadow over the remainder of the year. In this way, some seasons become defined by a single storm. The tornado is committed to memory for its size, ferocity, or for the trail it leaves behind. Manchester is one of them. Jarrell, Texas, is another.

But every once in a while, one storm becomes the yardstick against which all the others are measured. These are the storms of a chaser's dreams. This season holds just such an event. It will be unlike anything Tim has seen. It will be unlike anything *anyone* alive has ever seen. Through all the maddening busts, all the miraculous intercepts, all the close calls, Tim has spent years searching the plains for the ultimate storm. Now it's here.

PART THREE

CHAPTER NINETEEN

CHASE NIRVANA

FOR THE FIRST time since 2011, a small TWISTEX team assembles at the Samaras compound in Bennett, Colorado. On May 18, 2013, Tim, Carl, Paul, and Ed Grubb are gearing up to hunt down the swirling winds. In a way peculiar to storm chasers, their departure for the plains feels like a homecoming, even though they have no fixed destination. They haven't seen much of each other lately, but as the men pull away from Bennett in the GMC, it's as if the last TWISTEX mission were only yesterday.

They drive east into central Kansas and arrive in time to see a beautiful tornado near the town of Rozel. Storm chasers are nothing if not connoisseurs—they fawn over size, symmetry, clear contrast, and clean lines—and this tornado has it all. With the sun setting in the background, a sliver of its trailing edge shines cantaloupe orange; the remainder of its bulk reflects a bone-gray calcite. The structure is enough to leave the group in awe. From the cloud base, it descends as straight as a stovepipe, bulging near the center before narrowing gradually at first, then sharply near the surface. The tip finally vanishes behind a churning bowl of tilled sod.

Grubb sits on the truck's window frame and films over the roof, his camera jouncing as they cruise through the farm country. Occasionally, the wind sock on Paul's camera whipsaws into view. They pursue for half an hour, until the funnel lifts from the fields and retreats back into the clouds, like the tapering tail of a bull snake receding into tall grass. "There it goes," Tim shouts over the idling diesel. The wind begins to ease, and they hear the silvery sound of birdsong. It has been the perfect tornado, the perfect chase: no cities, no towns, no deaths.

For the first time since 2008, the only cameras in the truck are their own. There are no producers telling Tim what to do or where to go. And there are no hot microphones attached to his shirt. He is free to say whatever he wants without fear that the words will come to haunt him later on prime-time cable television. It's like the early years, before money and fame complicated everything. They trade jokes and barbs. This is the most fun they've had together in a long time. Even Paul seems more extroverted than usual, wisecracking with the other guys.

That night, the four check into a Comfort Suites in Pratt, Kansas, and find that they are sharing the motel with some old friends. Marc Austin, a forecaster at the National Weather Center in Norman, Oklahoma, and his wife, Sharon, had also chased the tornado near Rozel. Tim goes to bed early, as is his custom, but Paul, Carl, and Grubb file into Marc and Sharon's cramped room and drink beer late into the night. The day had been the kind chasers live for, and everyone is in high spirits. The usual indicators had been equivocal—the forecast didn't scream long-lived, significant twister—yet they'd all pulled it off, plucking a diamond from the rough. "It's just a tornado out in the field, and it's beautiful," Marc says. "It's kind of chase nirvana, that day was."

They pose for a photo of the five of them pointing dramatically at some imaginary off-frame threat, a goofy homage to Carl's signature stance on *Storm Chasers*, when he'd stare resolutely into the middle distance and level an index finger at the oncoming tornado. "[Carl] had gone through a tough time because *Storm Chasers* was done. He looked down," Sharon Austin says. But tonight, it seems the old Carl has resurfaced. She suspects chasing with Tim has lifted his spirits. The season is off to a raring start, and Carl is again on the road with friends.

There's an intimate proximity to chasing across states with the same group, in the same vehicle, that one seldom experiences in day-to-day life. A sense of camaraderie inevitably emerges over the long chase, along with a singular miasma. As Tim describes it: "You spend three or four days in a vehicle, it develops a certain scent. After a three- to four-day trip, we've got a lot of knowledge of each other and what's going on in our lives. Fortunately, our group gets along really well. That's the secret—you need people who are compatible. And of course, showering regularly is a good thing, too."

As good as it feels to have the gang back together, there are still moments early in the season that unearth an uneasy dissonance between Carl's style and Tim's. The Discovery Channel producers may have brought it to the fore, but Carl has always been the more aggressive chaser. A few days later, by May 20, they have gone south after a storm near Bray, Oklahoma, east of Lawton. Carl steers the probe truck along a narrow county road, hemmed in on either side with thick scrub brush and stunted trees. Tim glances down at an iPad, and his fingers move over the road map on the screen. He leans forward in his seat and peers up at the sky, eyes searching for signs of rapid cloud movement. A light rain taps against the windshield, punctuated by the heavier thwack of hailstones.

There is a tornado somewhere off to the left, behind the vegetation—that much he knows for certain. But they have lost sight of it. Worse, Tim suspects the vortex is beginning to rope. As a tornado becomes dislodged from the updraft, the powerful blowing cold of the forward-flank downdraft could send the dying twister careening into their path.

"We don't want to get right beside it," Tim advises Carl. "Let's be careful because we're in limited visibility."

The trees and brush form a dense screen; they're driving in the blind. But Carl doesn't let off the gas.

"*Careful!* It's right next to us," Tim says, more forcefully this time. "Slow down a little bit. Let's find out where it's at."

"It's right here," Grubb says, finally catching sight of the funnel.

"Yeah," Tim says, a note of exasperation in his voice. "I *know.*"

The contrast against the dull pewter sky is poor, but they can just make out the thin outline. Finally, they clear the trees obstructing their view and see the sky swarming with dark objects—brush, limbs, trunks hurtling end over end. The tornado is moving toward the road at an oblique angle, shearing through a stand of bur oak no more than a couple hundred yards off. They watch pieces of the trees lift, hover for a moment, and fall back to earth. Leaves and twigs drift onto the truck's rough Line-X coating like confetti. It's a sign they're far too close.

After the tornado dissipates, and the chase ends, Carl announces, "Wow! That was pretty exciting."

Tim, however, is brooding over the deadly course the day might have taken. In all likelihood, Paul's presence inclines him further toward caution. In previous seasons, Paul had typically followed at a safe remove in one of the mesonet cars. With Discovery Channel cameramen in the truck, there simply hadn't been enough room for him. But now he rides with Tim, and Kathy's words of warning still resound in her husband's ears. In the background of video footage taken by Paul, Tim is candid with Carl: "I didn't mean to get nasty, but if you get alongside it during the rope-out stage, we can get in a lot of trouble. I wasn't comfortable with it."

Carl's response is odd. Either he hasn't understood, or he is deflecting Tim's aggravation. He replies nonchalantly, "No worries."

After that, neither of them speaks for a while.

Any grievance must seem trivial later that afternoon when they hear about what happened in Moore, a suburb south of Oklahoma City. At the time, they had known only what they saw on radar: a nightmare velocity couplet over a densely populated city. As they drive south into Texas, the consequence of that radar signature reveals itself. Everyone in the car is glued either to a smartphone or a laptop. Two elementary schools, they learn, have been destroyed, and seven children are dead. The death toll has risen to twenty-four in all. More than 1,200 structures—houses, businesses, hospitals—will have to be torn down.

The tornado took a path through town that is eerily similar to the one sampled by Josh Wurman in 1999. Because even well-anchored homes have been swept from their foundations, the National Weather Service bestows an EF5 rating. While its forecasters had been clear about the potential for "strong tornadoes" today, what happened in Moore was no foregone conclusion. Most chasers—and forecasters—had thought the big show would take place farther south, around where Tim, Carl, and Paul were chasing. But something unforeseen transformed the supercell heading for Moore: a chance encounter produced a lethal outcome.

It was only after merging with another decaying thunderstorm that the Moore supercell kicked into overdrive and began cutting a trail of EF4 and EF5 damage. The question on every scientist's mind now is whether the merger touched off the storm's drastic intensification. And if it did, how could such a unique and seemingly random series of events be predicted?

Once again, the storm has shown mankind—Tim Samaras included—that there is still much to learn. The message from the sky today hadn't screamed killer tornado. But a middling storm just happened to collide with another thundershower's outflow surge, and now twenty-four are dead.

That night, the crew busts on a storm near Wichita Falls, Texas, then works its way east to book rooms at a motel in Sherman, just south of the Red River. The next day, they pass through Dallas, heading southeast after a possible target. Along the way, they see plenty of rain and hail, but no tornadoes. The following morning, May 22, they begin the long journey back to Colorado. Tim, Carl, and Paul need to prepare the LIV for the lightning project, which will kick off in a few days. Grubb won't join them for the next chase; he doesn't want to miss his daughter's birthday.

They drive north up Interstate 35, through Texas and into Oklahoma—a route that will take them right through the middle of Moore. By the time they pass Robinson Avenue and enter the damage path, traffic has slowed to a crawl. It is quiet inside the truck. Paul angles his camera out the window and shoots video; Carl snaps photographs.

They see trees that have been filed down to bare trunks, hung with metal and plastic. The mud-blasted vehicles piled along the highway look as though they've come from a scrap yard. As they pass through a neighborhood, the rows of houses stop all of a sudden. There is nothing on the ground but debris-strewn slabs and driveways that lead nowhere. After a few hundred yards, the houses begin again, practically pristine.

On the other side of the highway, they are stunned by the damage to the Moore Medical Center, an ostensibly well-engineered structure made of precast concrete and reinforcing steel. Even from I-35, they can see that the building's envelope has been stripped away where it faced the tornado. Part of its roof system has been peeled open. If they could have gotten closer, they would have noticed that a large metal Dumpster has come to rest *atop* the building. TWISTEX's resident EMT, Ben McMillan, had been present at the scene, helping pull people from a collapsed office building nearby as the tornado receded in the distance.

As often as those in the truck have had to pull up to a town reduced to rubble, the sight never fails to induce a profound melancholy. This is the storm chaser's moral conundrum: they come to see the fastest wind on the planet, but they know full well it may fall like the sword of Damocles on people's lives, homes, entire towns. Every chaser responds differently. If they believe in God, they pray for the dead and the living. If they're engineers, they may filter the destruction through the measurable: structural anchoring (or the lack thereof), debris impacts, and the cold calculus of wind loading. Some chasers dive headlong into the wreckage if first responders haven't yet arrived, the role of watcher subsumed by duty as a human.

These days will tear open anyone's world. They've often motivated Tim to push harder, to gather the kind of data that would have allowed even one more soul to survive. If the death sweeping past his window has such an effect today, he keeps it inside. No one in the truck is saying much now.

By the time they reach the end of the damage path, traffic begins to pick up again. They drive through the night and arrive in Bennett in the early-morning hours of May 23. Tim has a brief window—a few days at most—to ready himself for a different kind of chase. This isn't like loading the probe truck with a mesonet rack, a TOWER, and bunch of turtles. He'll be hauling into the field the most advanced mobile lightning observatory ever built. Not only must he double-check the camera systems and ready the box truck for a long haul, he's got laundry to do and another suitcase to pack. Since the interlude will be short, Carl stays on in the guest bedroom.

He's a good guest all around. Carl cooks like a gourmand and always insists on doing the dishes afterward. He is unfailingly kind to Paul and just as obsessed with movies. In the afternoons, they usually abscond into town and see the latest films, especially science fiction. This is one of the upsides of Tim's chasing, and what Kathy enjoys most: getting to know kids from a school in the Iowa farm country, or a guy from Lake Tahoe whom she might otherwise never have met without the passion that connects him to her husband.

But on May 26, there's no time to linger. The house is a flurry of activity. Tim is always a little stressed on the day of a mission. He wakes early and brews a pot of coffee. Periodically, he and Carl huddle over a laptop and discuss the weather models' various predictions. They intend to head out today and reach central Kansas by nightfall. He and Carl haul camera equipment out to the LIV and fill coolers with drinks and snacks.

When all is ready, Tim pulls around to the front of the house with a Chevy Cobalt, which he plans to use as a secondary chase vehicle, just in case they decide to go after a tornado. In addition to the funding he's receiving through PhOCAL, the National Geographic Society is underwriting his study of lightning and tornadoes inside plains "superstorms." To that end, Tim stows three probes in the sedan's trunk. Normally, he'd never attempt a deployment in one of the Cobalts, but with limited funding, Tim must save where he can—and the probe truck guzzles diesel.

At around noon, he kisses Kathy good-bye. He and Paul step into the LIV, and Carl mans the Cobalt. They pull down the long drive, and through the gate. The white van and the mesonet car turn north, heading toward the interstate. With the Bennett house's broad vistas, one can watch the vehicles recede into the distance until they are tiny specks. Eventually, they disappear from view.

Their farewell is a moment Kathy will remember. This is the last time she will see her husband and son alive.

————

Throughout the next week, Tim and his companions chase during the daylight hours and document lightning by night. The schedule makes for some exceptionally long days. "You're sleeping or you're working," says Walt Lyons, Tim's project partner. "These are twelve- to eighteen-hour shifts. It's not unusual for field campaigns to burn the candle at both ends. You're at the mercy of the storm."

On May 29, they settle around Salina, Kansas, as a staging ground for Tim's PhOCAL mission. That night, the many eyes of the LIV are fixed on a field of wind turbines like ancient cenotaphs in the strobing light. Reaching 12,000 frames per second, Tim's Phantom cameras slow the world down to an almost unfathomable timescale. An incandescent filament erupts from the top of the turbine, bisects itself, and blooms outward in delicate, arcing points of Day-Glo light. It all happens within microseconds, captured for the first time by the LIV.

By the thirtieth, the storms in central Kansas are withering. The men head south to Oklahoma. On a country road east of Guthrie, Tim and the guys are parked at the dirt entrance of a pasture gate, when a familiar crew pulls up behind them. Bruce Lee and Cathy Finley are out chasing for the pure enjoyment of it—no mesonets, no grad students. They'd seen Tim's Cobalt earlier in the week, and they're eager to reconnect now.

Carl stands on the side of the road, watching weakening storms move east. "We killed it," he calls out to Lee and Finley, referring to the day's guttering tornado potential.

"I thought we killed it." Finley laughs.

They set to swapping post-op reports on a fairly uneventful afternoon. Lee and Finley had seen a brief tornado. Tim, Carl, and Paul had busted. Finley shows Carl some shots she'd taken of an EF4 in Kansas two days earlier. He's distraught; the lightning project had forced them to miss the storm. Talk now turns to TWISTEX. Tim has scraped together enough money for a limited mission in June, just to test and calibrate equipment. Lee and Finley have updated the mesonets, and Tim is currently developing the next-generation TOWER. It isn't ready just yet, but he should have more time to work on the device as soon as the lightning project wraps up for the year.

They all lean against their cars and gaze out across the low hills and the post oaks. The sun is setting, and the rear flank of the storm is ablaze with its deep vermilion light. It feels like old times for a moment. But soon, Tim has to head back north. He left the LIV in Kansas to chase tornadoes in the Chevy Cobalt, and he needs to retrieve the rig.

In all likelihood, they'll be back down here tomorrow. The signs are pointing to a significant event in central Oklahoma. Lee and Finley are heading back to Minnesota, though; they're loath to chase storms near cities. They wish the boys happy hunting and begin to walk to their car. Dusk is coming on.

Carl calls after them, "See you in June."

———

Later in the evening, Lyons and Tim coordinate tomorrow's LIV deployment over the phone. Unfortunately, the best vantage point from which to observe transient luminous events will be somewhere to the north, far from the supercell action. If Tim chases tornadoes in central Oklahoma in the day, he'll be cutting it close. "If you go somewhere around Woodward," Lyons says, "get set up by sunset. This will be an early event. As soon as it's dark, I want you to be rolling."

The next morning, Tim steers south out of Kansas toward Oklahoma. Lyons speaks with him again at around 11:00 a.m., and by all indications he is bound for the planned destination. But along the way the three stop off in Alva, some sixty-five miles northeast of Woodward, and park the LIV in a lot near the Woods County Courthouse.

Eleven days after the horror in Moore, it has become evident that the Oklahoma City area is in for another hard day. Tim and Carl are astute forecasters. Apart from the environmental indicators that practically scream EF5 to even the newbie chaser, they would have noticed the oppressive air, sopping with moisture and heavy with heat. It *feels* like a historic storm day. The pent-up energy is tangible, visible.

Tim and Carl, storm junkies that they are, can't resist. They don't want to sit on the sidelines as they did at Moore. They decide to leave the LIV in the courthouse parking lot and drive south to intercept the storm. "Lightning by night, and tornadoes by day, was the way he had it set up," Lyons says. "I'm sure his intent was to be back in Woodward by the time it was dark."

He wouldn't leave the LIV some 130 miles to the north just to chase any old storm. Something epic is coming to the oil and cattle country west of Oklahoma City. The turtles are stowed in the trunk for just such an occasion, and Tim won't miss the big show.

CHAPTER TWENTY

A SHIFT IN THE WIND

A T AROUND FOUR that afternoon, the mesonet stations in the counties west of Oklahoma City begin to register the change Rick Smith has been dreading all day. The omen isn't anything as dramatic as a bloodred sunrise or a morning darkened by coagulating clouds. So subtle is the shift that those residing in Canadian and Caddo Counties are unlikely to notice it at all.

What troubles Smith, the warning-coordination meteorologist at the National Weather Center in Norman, is that the wind out of the south-southwest has deviated thirty degrees: and it's now coursing in from the southeast. This means that the dry line has advanced toward the metro area. Air masses from different directions are now converging.

Smith had hoped beyond reason that at least one of the pieces would fail to fall into place today. But the winds are "backing" now, summoning up the low-level convergence required for tornadoes. Events have been set in motion.

Here we go, he says to himself.

Smith had sensed the storm potential as soon as he stepped outside this morning. A miasmal haze dimmed the sky. The air felt syrupy. A native Tennessean and twenty-year veteran of the weather service, Smith made sure the tornado shelter in his garage was unlocked. He cleared the spiderwebs out and checked the batteries on the flashlights and weather radio. He gave his family a strict curfew: be home no later than 4:00 p.m.

By 9:00 a.m., he arrived at the National Weather Center, located at the southern edge of the University of Oklahoma. He made his way to the forecast office on the second floor, passing a cluster of desks arranged in a horseshoe pattern, and wall-to-wall windows with a view to the western sky. Each desk was crowded with as many as three monitors, displaying feeds from visible satellite and radar; the projections of short- and long-range computer models; and the office's various social media accounts, from which advisories are disseminated to the public. At the front of the room, a bank of flatscreen televisions was tuned to the local news.

Smith's first task was to convene the "morning huddle," briefing an already wrung-out crew on the day's dreadful possibility. They had all been pulling long shifts, issuing forecasts and warnings they hoped the people of Oklahoma would heed. They had watched in real time on May 20 as the radar signature over Moore killed neighbors in primary shades of red, green, and blue. The tornado passed within four miles of Smith's home and his family.

Among the forecasters on duty was Marc Austin, the chaser who'd celebrated the Rozel tornado at the Comfort Suites with Carl and Paul no more than two weeks ago. That glimmer of chase nirvana was already fully eclipsed, though. Yesterday, Austin had worked a nerve-racking twelve-hour shift on a severe-weather threat that had fizzled at the last moment. He was exhausted.

The winds *hadn't* backed yesterday. But by the morning of May 31, few of the forecasters believed the luck would hold.

In the huddle, Austin described the instability as "maxed out."

Once storms began to ignite along the dry line, they would go tornadic in a hurry, provided there was wind shear at lower levels. Without shear near the ground—without those backing winds—no vortex would be possible.

Tornadoes don't thrive on unidirectional harmony. They feed on convergence, on collision and opposition. This requires a spiral staircase of diverging wind vectors—a pathway for vorticity. It starts with a southeasterly wind near the surface, shifting southwesterly the higher it goes, and finally westerly once it hits the jet stream several miles up. As of this morning, one puzzle piece was missing. But Smith and Austin knew the winds would likely back to the southeast as the day wore on.

If their worst fears are realized, and this storm is as bad as they think it will be, the event should begin west of Oklahoma City and work its way east, reaching peak intensity during rush hour. That would mean the storm should mature by the time it enters the metro area on a day when softball teams from all over the country are arriving at the ASA Hall of Fame Stadium for the Women's College World Series. Hundreds of relief workers from the Red Cross and other agencies are still assisting with the recovery effort in Moore.

The question before the morning huddle was how to convey the gravity of this day to a populace still traumatized by the Moore tornado without inciting panic. "Are we going to break the glass and pull out the scary words?" Smith had asked. He knew what central Oklahoma had already been through this month. "That's all they've seen on the news: tornado damage, tornado debris. Pictures and video of funerals for the deceased. There was a saturation of information and a palpable sense of dread in the community."

Over the years, and most recently in Moore, he and his colleagues had noticed a dispiriting pattern during high-end tornado events. Some residents waited until the last moment and attempted to flee down the surface streets and highways. But by then it was usually too late. What they found was gridlock. People were dying in their cars.

Smith had decided to test a new strategy today. In the early afternoon, the Oklahoma Department of Transportation programmed its electronic highway signage to alert metro-area motorists to the coming storm, advising them to keep off the roads after 4:00 p.m. Though the official weather service policy is to recommend against driving in severe weather, today the office decided to urge residents who intend to evacuate to do so early, before the warnings are issued.

At times, it feels as though they are screaming into a void. How many will listen, and how many will die because they don't? Smith has been in this business long enough to know that people don't react to abstract threats. "Some of that is just human nature and how people respond to weather emergencies," he says. " 'I've got to see it when I look out the window or hear my favorite TV meteorologist saying the same thing.' "

Given the timing of the storms, dinner will probably be out of the question, so Smith had left the office a little after noon to grab a bite while he still could. It was surreal, driving through town toward Chick-fil-A, seeing all these people going about their lives as if this were just another day. He wondered which ones were visiting from somewhere else. Did they know what was coming in a few short hours? Did they understand that a fuse had been lit? For warning forecasters, there's a corrosive, low-grade anxiety to the wait. In the morning the office is abuzz with activity as the forecasters hustle to nail down the event, its location, time frame, and potential strength. After that, it gets quiet, even a little boring, which is another kind of stress. They're waiting for a bomb to go off.

The rest of the morning and afternoon had passed uneventfully. Smith set his plan into motion, liaising with local emergency managers, broadcast-news, and transportation officials. Austin tended to his routine duties, issued the aviation forecast, and assisted with graphics for release through social media. The office typically launches two weather balloons every day, one at 5:00 a.m. and another at 5:00 p.m. But shortly before two that afternoon, a special weather balloon rose at about a thousand feet per minute above the National Weather Center, a sprawling, nine-story structure of brick, precast stone, metal paneling, and glass-curtain walls. As the balloon buffeted along switchbacking rivers of wind, the radio signals were received by the tracking antenna and transmitted to the forecast office. As expected, the moisture had deepened in the lower levels. The layer of warm, dry air, known as the capping inversion, continued to keep a lid on the unstable mass below. That wasn't a good thing. The longer the lid remains, the longer these cubic miles of potential energy boil beneath a late-spring sun. It's like a sealed-off and smoldering room, the heat building and building inside. Open a door or a window—break the cap—and the room explodes. Heat shimmer becomes fire; unstable air becomes the storm. Austin was confident the sky would find its release at some point. But the longer he waited, the more he worried. All he could do was monitor satellite and radar for the first congealing cloud mass, the first echo.

By now, at about 4:00 p.m., they know it won't be long. The signal has arrived. The winds have finally backed. Visible satellite tells the story. For much of the afternoon, thin bands of cumuliform clouds had striped central Oklahoma like fish scales. Now they begin to coalesce into thick, lumpy braids, the towering cumulus throwing a distinctive shade onto the plains below. It's like watching the rapid growth of bacteria in a petri dish. Along the cold front and the dry line, which halves the state and trails into eastern Kansas, a string of storm anvils begins to swell, first in the far northeast of Oklahoma, now working its way down to the southwest.

The first radar echo—a storm tall enough to catch the radar beam—comes from Kay County, on the Kansas line. More follow like catching fire. By 4:46 p.m., Smith's office issues severe-thunderstorm warnings for Custer, Kingfisher, Caddo, Washita, Blaine, and Canadian Counties. The pace of storm growth is startling—a few puffy cumulus clouds have become thunderstorms with tops rising more than 50,000 feet into the atmosphere. "Think about a cloud potentially ten miles in height that wasn't there fifteen or thirty minutes ago. You can't really see that at the ground, but some of these updrafts are well over one hundred miles per hour. Things are going on in the atmosphere in such a short amount of time that it's hard to wrap your mind around," Smith says.

At 5:19, Austin observes the first evidence of a hook echo on radar in southern Canadian County. This does not necessarily mean a tornado is on the ground. But it means that the storm is beginning to organize, to rotate, and to wrap itself in precipitation. He is looking for evidence of continuity in movement, from the upper levels of the storm down as far as the radar's beam will reach. The sign he is searching for is convergent winds near the surface.

The forecasters glance up at the bank of flatscreens, where local news stations display live feeds from helicopters with eyes on the storm. They look for the shape of the wall cloud. Austin glances back down at the radar screen. He's searching for evidence that the precipitation detected by radar is moving in different directions in close proximity. That means rotation. If he sees a ball of precipitation at low elevations that's red on one side and green on the other, that's the signal.

At about 5:35 p.m., he spots it—clear evidence of tornadic rotation. Shortly thereafter, the forecast office issues the evening's first tornado warning.

"It's strange," Smith says. "Once the storms get going and you begin to see what may happen in more concrete detail, there's this feeling. You never quit trying to put information out, but at some point if you haven't done something by now—if you live in a mobile home and you haven't driven to a shelter by now—it may be too late."

CHAPTER TWENTY-ONE

EL RENO, OKLAHOMA

A WHITE CHEVY Cobalt pulls off to the side of Calumet Road, just south of the on- and off-ramps to Interstate 40, on an isolated stretch west of Oklahoma City. Behind the wheel, Carl Young films and shoots still photographs with a DSLR camera. Tim Samaras is in the front passenger seat next to him, and his son Paul sits in the rear. They peer through the windows to the southwestern sky at a mammoth Oklahoma supercell. It looks every bit as though it intends to pick up the deadly trail scythed through Moore eleven days before. This is a high-precipitation superstorm, characterized by sheer enormity and power rather than clean lines and discrete, graceful funnels. Its troposphere-piercing anvil looms nearly twice the height of K2. The mesocyclone and the wall cloud beneath spin low and close to an antenna tower's warning lights, darkening miles of prairie.

The wait can be enervating, especially when a chaser suspects what's coming. There's a helplessness in this foreknowledge. He can only keep his vigil, torn between two competing hopes: that a man who has seen so much will see something new today, and that when the tornado finally comes, no one dies. This close to Oklahoma City, though, he may be asking for too much.

The dire language out of the Storm Prediction Center is almost unprecedented. The latest sounding has been described as a "loaded-gun profile." CAPE values of instability are consistent with the highest ever recorded by the National Weather Center this time of year. The Bowdle, South Dakota, tornado of 2010 is likely the most powerful Tim has ever personally witnessed. Its CAPE values had maxed out just shy of 5,000 joules per kilogram. The values this afternoon are astonishing, rising in excess of 5,000 to 5,500 joules per kilogram.

On any other day, Tim would be his usual conscientious self, ever mindful of obligations. Lightning should be his primary mission right now. But he seems to have resigned himself to missing out on the most promising electrical display so far this season. At nearly six in the evening, it is unlikely he will make it back to Alva to collect the LIV in time. He'll pass on the once-in-a-season electrical event for a supercell setup that may never repeat in his lifetime. He's a tornado hunter at heart.

The mystery of what form will emerge from the storm is essential to the anxious wait. How does one patiently pass the time when, as a friend of Tim's once put it, you're waiting to see "the hand of God"? The coming cataclysm could resemble the awful thing that swept out of the pine curtain like a haze of coal ash in Dixie Alley two years before. It could look like the broad wedges of Manchester or Bowdle. Or it could look like no tornado he's ever seen.

Aside from their proximity to an urban center, this chase won't be as bad as Dixie Alley. Tim must take note of the country roads, which are laid out in a grid, albeit an uneven one. The prairie is largely level, apart from some low hills. Scattered over the green terraced ridges are oil-well-pad sites on graded squares of yellow-red gravel and sandy loam. The wheat and hay country is seamed with hackberry- and oaklined draws. There are gullies that bleed into freshets, freshets that merge into creeks, and creeks that branch toward the kinked trunk of the Canadian River—the only real natural barrier they might have to contend with.

Setting up on the southern cell is the obvious call, and it should be clear by now to Tim, Carl, and Paul that they are tracking the right storm. The sun isn't due to set for another few hours, but here, under the spreading anvil of the cumulonimbus, dusk is coming on fast.

At 5:41 p.m., Carl pulls back onto Calumet Road and drives south in no particular hurry. Excited chatter fills the cab. The distended wall cloud pulses with life. It's a cocoon in pupation, promising the emergence of another life-form entirely. After less than a mile and a half, faded asphalt transitions into gravel, and the circulation from their vantage becomes rain wrapped. In all likelihood, Tim and Carl have decided that by taking the first east road they come to, they can get out ahead of the rain and improve their sight lines.

Over the next twenty minutes they cover roughly three miles. At this leisurely rate, they probably stop to take pictures. Paul may be filming the evolving mesocyclone. Tim and Carl likely discuss chase strategy and the indicators of storm development. Smith Road is a straight shot east, across humped ridges and shallow creek bottoms, past the ivory spires of natural-gas-condensate towers and tank batteries. After a couple of miles, they leave the erosive cuts of the river valley, and the country before them levels out. The trees thin and give way to flat-topped blond wheat ready for the combine. The first cloud-to-ground strokes luminesce through deepening slugs of precipitation. It won't be long now.

At 6:00 p.m., still traveling east, Carl begins to film again at the intersection of Smith and Heaston Road. He holds a camera in one hand and the wheel in the other. Tim's face appears in profile, with that unmistakably hawkish nose and thick-framed glasses. He is staring toward the south, down Heaston, the storm no more than a mile and a half distant.

Ashen braids of cloud race over the prairie from north to south, looking low enough to touch in places. If this entire circulation pool comes down—if its size is any indication of what's to come—this is going to be quite a chase. Tim knows the chrysalis of a legendary monster when he sees one. "Oh, my God," he says. "This is going to be a huge wedge."

Events will soon move quickly, and they won't slow until the end. The last few moments before a killer tornado touches down are like the breath drawn before the plunge. The beginnings of the funnel are often beautiful and evocative—a flaring and bunching of cloud material, like the dance of a murmuration of starlings across the sky. But all is moving at a faster speed this afternoon.

Within seconds of the first wisps, there is a tornado on the ground, a broad cone out of which thin vortices lash and spear the prairie. A few rapid strokes of the Cobalt's windshield wipers pass, and a series of oscillations are transmitted along the length of the funnel. It quadruples in width. The growth is so sudden it can seem as though a distortion is at work, a visual artifact produced by a lens. But it should serve as a reminder that what they chase isn't bound to some visible spectra. It's the closest thing to a ghost in the physical world. "Wow," Carl exclaims. "My God . . . Look at the tornado! Just to our south."

They continue east, driving forty and fifty miles per hour as the sedan shudders over the washboard dirt road. Tim and Carl would expect the tornado to close in on them gradually, taking a diagonal northeast track that would place it on a collision course with El Reno. Instead, it seems to be receding, falling away ever farther to their southwest. Smith Road is no longer a viable intercept route. It is time to readjust and close the distance. At about 6:06 p.m., the Cobalt slows and turns south down Brandley Road.

The shape on the southwestern horizon is alien, otherworldly, a plume of smoke that lives and breathes, swirling of its own accord.

"It's heading straight for Oklahoma City," Tim says.

They continue south for a mile, but Tim prefers to operate with a buffer, and if his choice to turn at the next intersection is any indication, he's sensing that theirs grows threadbare. Turning east on Jensen Road means caution, safety.

Yet with each moment they drive eastward, the tornado veers farther out of range, a behavior they must find bafflingly aberrant. Most
mature twisters follow the parent storm and the prevailing winds from
the southwest to the northeast. This one is clearly tracking to the southeast. The ideal intercept strategy would be to stay north, get ahead,
drop down a little south if need be, and deploy. They're in perfect position for a conventional, northeastbound tornado. But this is clearly not
a conventional tornado. Today, they're barely keeping up.

Regaining their edge will require some white-knuckle maneuver-
ing. They'll have to take the next south road and push close enough
to keep the tornado in range, but not too close. There's another east
road about a mile down, which should keep them clear of the outer
circulation.

Just before 6:09, Tim, Carl, and Paul make their move. Carl steers
the Cobalt south down Chiles Road in an attempt to close the increasing distance. But the nearer they get to their east route, the clearer
their mistake becomes.

For roughly forty seconds, the car remains unusually quiet. Carl is
the first to break the silence. "Is the airport down another mile?" he
asks, though he already knows the answer.

Reuter Road, the planned east turn, is a dead end. For the most
part, the road network is laid out in a reliable square-mile grid. Not so
with Reuter, which terminates at the El Reno Regional Airport. This
means Tim, Carl, and Paul are now committed to driving an additional
mile to the south—much closer to the tornado than they would prefer.

By 6:10, as they pass Reuter, they are no more than a mile and a half
from the outer edge of the condensation funnel. Carl stammers about
spotting Reed Timmer, their former *Storm Chasers* costar, perhaps in an
attempt to make himself and the others feel as though their position
isn't quite so precarious. There are only two options available to them
at the moment, and neither is appealing. They can turn around and
backtrack to Jensen Road, then continue east; this will almost certainly
knock them off course and preclude an intercept. Or, they can grit

their teeth and continue south, courting the storm's northern flank as they make for the next east road, called Reno.

Tim and Carl choose to press on. At about the same time, the tornado's south-southeast track shifts. It is now moving almost due east. From their perspective, they would notice that the tornado is no longer off to their two o'clock—instead, it is directly ahead of them, sitting over Chiles Road, a sooty parabola the width of some eight football fields. They're still in the chase, but as it stands, they're being outmaneuvered.

———

At 6:04 p.m., and a little over three miles to the east of the Cobalt, Howie Bluestein and two graduate research assistants set up on a rise at the southern outskirts of El Reno. Nearby, his RaXPol mobile radar generator thrums as the antenna revolves atop its flatbed pedestal. Suncured wheat spikes sough and toss in the tremulous fields to the west, where the wind whips the crops into hypnotic waves. But Bluestein and his assistants are focused on a sight farther out.

Just south of due west, the storm's lowest level is hidden behind the gentle swells of the intervening miles, yet there is no mistaking the vertiginous plunge of the cloud line. "I'd say that's a tornado," says Jeff Snyder, a lanky Minnesotan and research scientist. The broad cone vortex is the color of blue gunmetal, and the cloud motion is startlingly energetic. The storm is gathering strength. The scientists can practically sense the velocities it is preparing to unleash.

Bluestein gives Snyder an instruction, though it is difficult to hear over the generator. "Do you want me to do tornado mode?" Snyder shouts, in reference to the antenna setting that would focus the radar beam closer to ground level.

"No, not yet," Bluestein replies. He notes that the tornado is still too far out, five to six miles away. Better for now to maintain a broader scan.

"I'm only doing seventy-five-meter range gates instead of thirty."

"Okay. How long will it take you to do thirty?"

"For me to get reset up, it's gonna be down for twenty to thirty seconds."

Bluestein can't afford to have RaXPol off-line, not with an intensifying tornado in progress. "Let it go," he says.

As the minutes pass, however, it becomes apparent the tornado is tracking to the east or southeast rather than to the northeast. This is a propitious development for the town of El Reno. But for the researchers, it means they will soon need to reposition. With the sun slipping down toward the western horizon, the once-crisp contrast fades into a brume of rain.

"Do you see it in there?" Snyder asks. Wind presses at his back, drawn into the storm as though its pull is gravitational. "This inflow is something else."

He glances at the data accumulating on the screen in the backseat of RaXPol, then looks to the real-time radar display, which reveals a prominent hook echo. It's still in there, of this there can be no doubt. Snyder can even make a crude estimate of its wind velocities: "It's folding over at least thirty . . . sixty, greater than sixty meters per second," he says. "That's over 140 miles per hour."

He's shocked by how intense the tornado has already become. Though he couldn't know this in real time, the true ground-relative velocities are closer to two hundred miles per hour.

By 6:15 p.m., Bluestein's assistants are retracting the hydraulic stabilizers and prepping RaXPol for the next move. They believe the tornado is still tracking to the southeast, which will take it farther and farther out of range.

They need to keep ahead of it and also avoid the large hail core that radar indicates is on its way. The storm will ultimately set a state record for hail size, producing chunks of ice up to six inches across. It's just one of many records to be set tonight.

As far as Bluestein is concerned, they have only two options: either they drop south after the tornado or head east, keeping the storm on their right side. After a brief deliberation, he judges it safer to drive east. The roads directly to the south aren't paved, and he can't risk getting stuck in this heavy rig. The Canadian River also presents a natural obstacle, with only a few bridges. There's a southbound highway farther to the west, but he isn't confident there will be enough time to make for it and cross ahead of the tornado.

Going east may mean that he will lose sight of it in the rain, but at least he knows he won't get hit. RaXPol pulls onto a deserted I-40 and motors east, Snyder pushing the diesel to the redline. He can only manage fifty-five miles per hour. The easterly headwinds have become so intense that the antenna not only behaves as a parachute, it stops rotating altogether.

As the team examines the scraps of radar imagery that do make it through, Bluestein notices something curious. In fact, he is positively confounded as they travel along the southern flank of El Reno, past the oil-field-supply depots and the livestock auction. The tornado is now *behind* them, when it should be somewhere off to the right.

How could this be? *If the tornado is moving southeast, and you move east, and suddenly the tornado is west of you,* he thinks, *what does that mean?* The most likely explanation, he concludes, is that the supercell is producing tornadoes cyclically. The vortex they had witnessed earlier must have roped, making way for the new twister, which would form farther to the north—behind RaXPol.

Only later, when Bluestein has had a chance to analyze the data, does he discover that he had been dead wrong. It wasn't a new tornado. It was the very same beast. Instead of drifting away from them, it had swerved dramatically northeast. "And I thought, *Oh, my God,*" Bluestein says. He and his crew were one decision away from losing their lives. Against such a strange track, venturing south would have brought them straight into the maw of the storm.

CHAPTER TWENTY-TWO

THE DRAGON'S TAIL

CARL BRAKES AS he approaches the intersection with Reno Road. At 6:12 p.m. they've finally arrived at their east route, and just in time. The tornado is roughly three-quarters of a mile away at their ten o'clock. They're close enough now for a view afforded to few others through the rain. The visible funnel is nearly nine hundred yards wide, but the monster is more than what the eye can perceive. Its unseen circulation, a 2.4-kilometer ring of Category 5 typhoon-strength wind, is no more than half a mile away. And it's closing. After its initial southeast trajectory, the tornado has started to edge a few degrees northward, heading due east now. They'll be skirting even closer than they'd planned.

To devote his full attention to the road ahead, Carl passes the DSLR to Tim. He turns left onto Reno, the tornado now visible out the passenger window.

An intercept is not yet out of the question, but Carl knows they will first have to gain on the tornado, pass it, and put some distance in between for the deployment. He needs to drive fast to keep their hopes alive, sixty miles per hour or more down a rain-slicked country road in a hard and veering wind.

Tim is unnerved by their speed, but not because he doubts Carl's ability to handle the sedan in these conditions—he's done so many times before. What frightens Tim is the specter of all that can go wrong farther down the line. The tornado, he knows, is like a steel tooth on a coil-spring bear trap. The question is not *whether* it will snap shut, but when. Once the rear-flank gust front coils around the updraft, it could dislocate the tornado and send it caroming into them, dying but no less deadly. Tim has seen it happen any number of times, and fears they are driving straight into the jaws of the trap.

"Okay, we've gotta be careful in case this thing wraps up. . . . I would slow up here, because if this thing starts moving to the north, we're in trouble," he says. "Slow up. . . . We're almost right alongside of it here. *Slow up!* Let the thing go off to the east a little bit . . . see if that thing transverses us."

Carl sees nothing but clear road ahead, and he believes they can't afford to lose any more ground. It is a familiar push and pull between Tim's relative caution and Carl's aggression. This is their opportunity to overtake the tornado, he insists.

But what Carl can't know is that the usual distinctions of vortex anatomy have become virtually meaningless. This is more like the storm Wurman encountered in Geary, Oklahoma, when the DOW became a probe. The entire mesocyclone has sunk to the ground, creating a tornado that's practically purpose-built to ensnare chasers with its hazy boundaries, torrential rains, and roving pockets of lethal wind.

Even with access to real-time mobile radar, Wurman strayed inside at Geary. Tim and Carl have no such advantage. In fact, it's apparent they don't know what they're looking at from one moment to the next. Is that merely rain, or is it the outer circulation? Is the visible funnel the full tornado, or simply a part of something much larger? How big *is* this thing? The problem is one of perspective. This is not the kind of tornado they think it is. It does not end where they think it ends. With nothing but their eyes and updates every five minutes from the weather service's stationary Doppler, their decisions aren't fully informed.

Shortly after they pass south of the airport, the current shifts suddenly. A stiff headwind out of the east wedges and pries at the hood of the sedan. Without realizing it, they have entered the outer circulation of the tornado.

Tim spies hanging motes of dark chaff in the sky all around them. "We've got debris in the air," he says. As if to drive home his point, something heavy glances off the Cobalt. "*That's* the problem."

Tim sets the camera down on the floorboard, where it remains for the rest of the journey, and he begins to chart a new course.

He instructs Carl to take Reformatory Road, the next north option, shortly before 6:15. Wind and debris continue to batter the sedan. As they reach the corner, it's unlikely they mark the alteration in vortex structure that occurs at the same moment.

Directly to their south, the storm is pouring rain onto the empty fields, heavier and heavier with each passing second, darkening to the point of opacity. Then, the absolute gray surrounding the tornado breaks and admits a weak light. Alternating bands of rain and wind-cleared air sweep toward the heart of the storm, one after the next. But one of the rain bands is different from the others, cohering, hardening; it isn't rain. It sweeps down out of the cloud base, dark as onyx, drifting over the fields like the slow lashing of a dragon's tail.

A careful observer would be able to see the birth of an embedded subvortex—a tornado within the tornado. But Tim, Carl, and Paul are likely too distracted by the unrelenting inflow and their precarious footing to take note. Behind them, the subvortex takes root and begins to trace a near-concentric orbit within the parent tornado. From the Cobalt, it would have been visible for only a moment, then seemingly reabsorbed. But it isn't done. Before long, it will return.

———

At 6:16 p.m., Josh Wurman levels DOW6 on a short paved drive off State Highway 4, where a steel gate opens onto broad pasturage to the east. The hydraulic outriggers cantilever the bright blue seventy-ton

International truck. The 250-kilowatt magnetron transmitter buzzes in the steel-fortified rear cabin. A communications mast periscopes fifty-six feet above, all of it giving the rig the appearance of a Transformer in metamorphosis. Roughly fifteen miles west, far beyond the low, single-story shingle roofs of a suburban Yukon subdivision, the tornado traces a snaking path.

Three miles to the southwest of DOW6, a second unit, the Rapid-Scan DOW, has taken up an ideal position in the farm country, with almost nothing taller than a stalk of wheat between the antenna and the storm. For sampling an eastward mover, the team is well situated, at least for the moment.

Wurman and his colleague Karen Kosiba should be on a plane to Helsinki right now for a conference, but neither could resist the beckoning volatility over the southern plains. Instead, they set out this morning from the Wichita, Kansas, airport with two DOWs and one mesonet vehicle, which members of their team use to deploy Wurman's pods. Three hours later, they arrived in Norman, Oklahoma. Given the possibility that the tornado would mature by the time it reached the outer-ring suburbs of Oklahoma City, they wanted to be prepared for a rare opportunity to document the relationship between damage and velocity in a population center, an event Wurman's radar has witnessed only once before. As the storms touched off, Wurman and his team worked their way through the suburbs southeast of here before setting up at the thinning edge of the metro area.

Now, the antenna rotating on its axis, Wurman's DOW6 has begun to scan. RSDOW follows suit two minutes later. Much of what follows will be discovered only after he has analyzed the radar data in detail.

At around 6:17, DOW6 detects near-concentric tornadic circulations, one 2,000 meters in diameter, and a second 150-meter subvortex nested almost directly within its core flow. This is what Tim, Carl, and Paul would have just been able to glimpse through the rain curtains. By 6:19, other vortices northwest of the parent tornado are ingested into the circulation. Then, a profound evolution takes place. At 6:20, the embedded subvortex becomes dislocated. It drifts from the center toward the farthest boundary of the circulation and embarks on an entirely novel orbit. "It's hard to know why it changed in structure, from something internal to something external," Wurman says. "You could speculate, but because it was such a difficult storm, nobody got the data they were hoping for. It was just too hard. There are a lot of questions that probably won't get answered."

Those questions will come later, however. For now, across the street from the Canyon Creek subdivision, Wurman peers into the display inside a cramped cabin, cooking in the radiative heat of high-voltage electronics. He knows only that the DOW is reading a baffling and nearly incomprehensible bolus of violent wind. He studies the screen with growing puzzlement, attempting to orient the other DOW and his pod-deployment team in time and space. Yet he can't discern the pattern. Where will the most violent winds show up a minute from now? Five minutes from now? Even with DOW6 scanning a full 360 degrees every seven seconds, he can't make sense of any of it. It will be impossible to direct the pod deployment team with any degree of safety. What's worse, it is clear the subvortex, once unleashed from its tight orbit within the main circulation, contains winds of deadly intensity—well over 200 miles per hour.

Wurman is not going to lead himself or his crew anywhere near velocities like that. "You'd better be careful with that kind of thing," he says, "because it could kill us." He calls off his pod deployment team.

CHAPTER TWENTY-THREE

THE CROSSING

Tim, Carl, and Paul hit the turn onto Reformatory Road just after 6:15. Rain now rips over the road in dense horizontal sheets. Objects plunge through the air like knots of sparrows. Carl rounds the corner and guns it north, with conditions pushing the Cobalt to its limits. Each second is precious. Whether or not they realize it, the tornado is lunging for them.

The moment they crossed into the core flow, with debris ringing off of the Cobalt's frame, the vortex snapped its jaws to the northeast, just as Tim had warned. Reformatory becomes an escape route. If not for their turn here, the vehicle would have been overtaken.

Even after Tim has predicted its next move, the hair's-width margin comes as a shock. This tornado is far larger, and moving far faster, than they could have expected.

The white Cobalt now shoulders through an inflow current like a boat struggling upriver. It's harrowing driving, but they're able to open up a buffer. Judging by the chatter in the car, they're relieved to be out ahead of the beast. They plot their next maneuvers: "Now we go up north," Tim says, "then east."

Distance, however, does not gain them perspective. The stiff southbound inflow winds are attended by a slug of rain winding centrifugally around the tornado's northern flank. What had once been so clear, so readily tracked, quickly dissolves into the gray. Only chasers to the south of the storm can see the tornado now.

As they make their way farther north and plot a new route, Tim's phone rings. It's a New York producer with whom he's been working, calling for an update at a particularly inopportune time. "Yeah, yeah. We're at . . . the tornado is about five hundred yards away. I really can't talk right now!" Tim says. "It's just south of El Reno. It's gonna be on the ground for a long time, and it's heading right for Oklahoma City."

Tim does his level best to end the phone call quickly, but in a burst of pique, it sounds as though he mutters "Goddamnit!" under his breath. He hangs up after forty-five seconds. It must seem like an eternity under the circumstances.

A mile up Reformatory Road, Carl swings east again onto Reuter. They're still angling for an intercept, even though they've been given every reason to abandon it. Between the untimely dead end at the airport, the perilous push to the south, and their brief penetration of the debris core, the chase has proven problematic from the first.

But Tim isn't the same chaser he was on that dirt road near Last Chance some twenty years ago, or even at Manchester. He's spent so much time in the presence of violent storms that the old pang of fear has dulled. And while he has often said that nothing scares him more than the tornado he can't see, he and Carl have penetrated the rain before not knowing what awaited them on the other side. No-man's-land is where Tim found his greatest victories. He has succeeded by toeing the line between danger and safety. And as he's gotten better, that instinctual border has drawn closer and closer to the storm's deadliest winds. He's broken or stretched chasing's cautionary rule so many times it no longer has the same hold. He is like the expert climber who knows he can succeed without a rope—who's felt each hold and scaled each pitch a hundred times. In the back of Tim's mind, he knows it can always get worse; but his instincts haven't warned him off today's chase.

He must be curious. He must want to see what is behind the rain. Tim has never encountered this kind of storm motion before. He has never seen anything that careens to the south, the east, then northeast. Perhaps it is the fog of the chase, but he seems to think they have turned a corner, and that their lot is about to improve. After miles of shuddering over dirt roads, Reuter transitions to asphalt. The vortex seems to be easing more easterly. Now that they've established a reasonable buffer, this is their chance to regain lost ground. After a year's hiatus, perhaps Tim thinks TWISTEX, even in its diminished state, will make history today.

Still, trouble hounds every providential development. For some 500 yards, they drive on in the blind, their view of the storm obstructed by a plains windbreak of thick red cedar. While Carl believes pavement will enable a highway pace, the north–south inflow buffeting the sedan continues to check his speed. The power lines to their right strum in the wind. Maintaining a straight heading requires his complete concentration. In half a mile, Reuter switches abruptly back to gravel. The Cobalt, built for city driving, not rally-style terrain, handles poorly on the dirt road, shimmying in the wind. In past TWISTEX missions, veteran operators of the mesonet sedans had refused to leave the pavement for this very reason.

Despite these conditions, Carl is determined to outrun whatever hides behind the rain, or at the very least to keep up. He is clocking between forty and fifty miles per hour. Tim seems ill at ease, at one point loudly alerting Carl to an approaching stop sign.

"I see it!" he replies, with a touch of irritation. The men go silent for a while.

The next time Tim speaks, it is to note that the funnel remains out of sight, concealed behind the rain curtain. Without context, an observer could conclude that the dark mass of cloud bears nothing more worrisome than a ruinous deluge for the local dryland wheat crop. At its leading edge, though, the chasers may spot a crescent-shaped penumbra of centrifuged water, lit up by the weak early evening sun.

Something is shedding that rain, sending it spiraling onto the plains ahead.

By 6:18, as they near the intersection with Choctaw Avenue, a blue Toyota Yaris pulls onto Reuter from the north, no more than fifty yards ahead. The sound of wind-driven rain hisses against the windshield in swelling and slackening volume. The Cobalt bucks over a railroad crossing and speeds another half mile east before entering the wooded bottomlands that line a spur of Sixmile Creek. Once again, the sight lines to the south are fitfully blinkered by a belt of cottonwood and hackberry. A minute later, US 81, a divided, four-lane highway, lurches into view. At the intersection, Carl brings the Cobalt to a halt behind the Yaris.

Earlier, they had discussed diving south here. If Tim and Carl consult the weather service's radar feed for evidence of the tornado's location, the data will be of limited utility now. The last update refreshed nearly five minutes earlier, about the time they wriggled free of the core flow. When they peer south down the highway shortly after 6:19 p.m., the radar's obsolescence becomes chillingly apparent; they are astonished to see that they have failed to outpace the tornado.

US 81 disappears some 400 yards to the south, as if it has terminated at the foot of a sheer crag, rising above the highway with neither grade nor foothill. It swallows the horizon for more than two miles in either direction. If their eyes could penetrate the shadow and rain and dust, they would see vehicles tumbling.

Turning south is obviously out of the question. And with the tornado moving off rapidly to the east—or is it the northeast again?—heading north would mean placing themselves hopelessly out of position for the intercept.

"So, this is the highway. . . ." Tim begins.

"Yeah."

"We're just gonna have to . . ."

"Keep on going," Carl finishes Tim's thought.

Disagree as they have in the past, this time Tim and Carl are in perfect accord about their next move. They are not giving up on the storm of their lives. To stay in the game, they must keep going east. The Cobalt crosses the highway's four empty lanes and picks up the dirt road on the other side.

———

Due south of the intersection, Mike Bettes, an on-camera meteorologist at the Weather Channel, adjusts his earpiece for the live shot. At 6:14 p.m., he's standing near the southbound lane of US 81. The mile-wide darkness is still west of the highway, and the light around him, refracted through an apricot sunset and deep blue rain, looks stained, like stagnant water steeped with the tannins of leaves. Bettes wears a bright blue TWC Windbreaker and a ball cap covering short blond hair. "Guys, that's it," he shouts. "Right there! There's the tornado. We are just north of Union City, south of El Reno, south of I-40. Cops have blocked off 81 at this point. They're not letting traffic go southbound because the tornado could go right over top of Union City, or pass very close to it.

"Look at that *monster*. This is a huge tornado. If you live in Union City, south of El Reno, you have to take shelter *now*! There is no more time to waste. The movement on this is going to be almost due east."

The satellite feed deteriorates for a moment, and the audio cuts out, but Bettes is describing the wrapping rains, his hand swooping and cutting illustratively. "We are probably roughly a mile away from it right now, and it's absolutely *enormous*! There is no time to waste. Right now you have to get to your shelter as quickly as possible.

"This thing is moving relatively slowly," Bettes says, addressing his crew as much as the live viewing audience. "What we may end up doing, guys—because we could be putting ourselves in a dangerous position here—we may actually quickly try to get into our cars and get south of here. That'd be the safest place to do that. I think we have to go now in order to stay ahead of this and not get run over by it."

At 6:16, the men pile into a satellite truck and two GMC Yukon XL SUVs. Bettes rides in the front passenger seat of the second GMC in the convoy, and Austin Anderson, a field producer from Texas, jumps behind the wheel. They pull onto US 81 and dash south, trying to cut ahead of the eastbound tornado. Without realizing it, by 6:19 they have penetrated its weak northern flank. Bettes was wrong; Tim, Carl, and Paul aren't the only chasers who have misjudged both its speed and its size. The blinding rain is the first sign the convoy is in trouble. The light through the windows fades and the sky darkens.

Cameraman Brad Reynolds rolls the window down and trains his lens just ahead and to his right. The trees lining the road lean north; limbs flagellate in the current. Then the line of trees ends and they get a clear view.

The tornado is just off the road, almost directly to the right and coming for them. In one second the subvortex is an opaque stovepipe. In the next, it breaks down and disperses into two, then three, and as many as four suction vortices, like geysers of pure vertical motion.

"Oh, shit," Bettes mutters. "Oh, shit."

We gotta get out of here, Anderson thinks. *We've got to get past them before they hit us.*

The road in front of them vanishes. "Hold on, brothers," Bettes shouts. "Hold on."

Reynolds rolls his window up, and the camera sees little through the tint but the red smear of passing brake lights.

"Everybody, duck. Go-go-go-go. Just keep it going if you can. Everybody, duck down."

Reynolds's window shatters.

There is the rushing sound of wind, a wall of white noise, and Anderson experiences a peculiar floating sensation.

Then, the SUV comes down hard and his memory goes dark.

When he comes to, Anderson's head is bouncing against the earth through the broken window each time the SUV rolls onto his side of the cab. Reynolds's camera picks up a confusion of tumbling and the grunts elicited by the impact of bodies against hard interior paneling. Then the camera is ejected from the car. The image is blurred, but the SUV can be seen rolling over and over, receding from view.

When the GMC finally comes to rest, on its tires, the wind surges through the cab and Anderson spits the gravel and soil from his mouth. Bettes shouts out to his team. He and Reynolds climb out and try to pry open Anderson's door, but it won't budge. They extract him instead through the other side. Anderson stands up and feels sharp swells of pain in his ribs with every breath. Though the adrenaline partially inures him to the extent of his injuries, he's cracked his sternum, crushed several vertebrosternal ribs, and fractured a vertebra in his neck.

The men look at the SUV. It has been dismantled—its luridly decaled windows shattered, side-impact air bags protuberant, the roof over the front seat nearly flush with the hood. They had been traveling south down US 81 and now find themselves in a field east of the *northbound* side. The SUV traveled across the broad median, then across two more highway lanes—a distance of some 200 yards. Others begin to emerge from overturned cars and trucks nearby. Two men, a volunteer school bus driver and an oil-field hand, had been hit roughly a mile to the southwest, a minute before the subvortex overtook the Weather Channel SUV. The bus driver, Billy O'Neal, is already dead. Dustin Heath Bridges, the oil-field worker, soon succumbs to his injuries.

Anderson can see the trailing edge of the tornado some 150 yards out. It's more than two miles wide now and heading northeast.

CHAPTER TWENTY-FOUR

THE LAST RIDE

CARL RACES ACROSS the four lanes of US 81 and plows into the hanging traces of grit that billow behind the Toyota Yaris in front. The trees thin and the red-bed plains slope gently before them. Carl's outlook brightens. The rain is easing and the road ahead appears dry.

It has cleared up enough that if they glance out Tim's window just after 6:20 p.m., they will sight the thin annulus and sod corona of a satellite vortex, no more than 250 yards distant.

After only eight seconds, though, it is ingested by what can only be described as an encroaching wall. Confusion begins to grip the men in the Cobalt. They have been flying down country roads at nearly fifty miles per hour, and they can't seem to gain an inch. Tim suspects the tornado is racing at forty miles per hour at least.

If Tim consults the latest radar scan from the nearest stationary Doppler, updated a minute before, he will note that it places the tornado core signature two miles to his southwest. But at this proximity he is likely chasing by sight alone. And this is just as well, because the radar is misleading. Much has changed in the last sixty seconds.

The tornado is steadily swallowing the distance. Just as the Cobalt passed US 81, the twister jagged to the northeast, even more severely than when they'd skirted the circulation just minutes ago. It isn't only that it's turning toward them, now; the tornado is expanding. It's *growing* toward them. They're not, as they'd thought, comfortably ahead; they're on the knife's edge.

Some thirty seconds later, the rain begins to fall with such ferocity that their windshield wipers can't clear the water. The world around them closes in. Visibility is reduced to mere tens of yards.

Their grip on the increasingly sodden road is deteriorating, and a seventy-mile-per-hour headwind out of the northeast further slows their progress. Given the amount of ground they have covered, Carl is still maintaining a quick pace despite the conditions, at least forty to fifty miles per hour.

Their eyes strain to the south, searching.

"Now I see it," Carl says, "well, maybe I don't."

"It's like it's just a bunch of rain here," Tim ventures.

Then, for the first time, he must begin to understand. There is no way of knowing exactly what Tim sees in the south, but the timbre of his voice shifts. It isn't panic so much as a dawning comprehension: *The tornado isn't gone; it's just enormous.*

"In fact, uh, keep going. This is a very bad spot."

With that, Carl's DSLR camera goes silent, having reached the capacity of its five-gigabyte storage disk.

The camera pointing out the rear window of storm chaser Dan Robinson's Toyota Yaris picks up where the audio recording leaves off.

From the moment the Cobalt crosses Highway 81, it is losing ground. The tornado sprints across the fields, its inflow battering the sedan. While Robinson pulls away from the gathering shadow, the darkness fastens itself to Tim, Carl, and Paul. The Chevy's headlights dim and flare with the passing of intervening rain curtains, winking from view behind the gentle heave and fall of the prairie. Within a minute, beneath the lowering vault of tungsten-colored cloud, the headlights dwindle to a single, small point of light.

The Cobalt is now experiencing headwinds upward of 110 miles per hour. Its four cylinders are not equal to the task of hauling three grown men and three steel probes over a muddy road. They can manage no more than twenty to thirty miles per hour.

As the distance between Robinson's camera and the storm grows, a dark wall fills the left edge of the frame. The vault lifts and the sub-vortex that has haunted the Cobalt's trail since Reformatory Road now reveals itself. You'd have to catch it between the distortions of sheeting water on the rear window, but it's there: a tower, black as shale and rising vertical from the earth, its base lapped by turbulent vortices.

Tim's car is now well within the primary tornadic circulation. The more powerful subvortex closes in. The headlights glimmer once more at about 6:22 p.m.

In all these years, Tim has learned to see the tics and patterns of the vortex. His probes aren't all that have entered the unknown, glimpsing places no one alive had ever seen before. Tim has as well. And at these moments of extremity, it has always been his talent to see when the door is closing. He has always been able to find the seam, and to slip through to safety.

But this time, it is too late.

This is the tornado he can't outrun.

The dark wall closes over the road, and the headlights are extinguished.

From the fleeing Yaris, this is the last living image of Tim Samaras, his son Paul, and Carl Young.

They have a little more than a minute left before the real killer—the subvortex—engulfs them. What they say to one another—whether there is time to say anything at all, or whether they even see it coming—remains a mystery.

But we do know what happens inside the storm, where no camera, no eye, can penetrate. Josh Wurman's DOWs are still scanning. At around the time the tornado's subvortex crosses US 81 and strikes the Weather Channel SUV, it becomes dislocated from its near-concentric loop around the core and is sent drifting outward. It begins to orbit the main tornado, describing a series of broad loops whose apexes manifest as tight curlicues, like the shapes a child might create with a Spirograph. Depending on its position, the velocity of the subvortex fluctuates drastically. Along the broad arc of the orbit, its speed increases exponentially as it races past the main circulation. But at each apex, the subvortex enters a tight curl that renders it nearly stationary, its high-speed winds scouring a single small area for seconds on end.

When the vortex reaches its first apex, Tim, Carl, and Paul are still within view behind Robinson. They have 130 seconds left. The subvortex arcs counterclockwise across the storm like the hand of some exquisite clockwork. Then it hooks rapidly across the southern rim of the tornado and slingshots to the northeast. More than 200 yards in width, it screams over the fields at up to eighty miles per hour. It sweeps toward zero hour as though it has been counting down.

With the storm moving over them from out of the south, the east is the last direction to which Tim would look for death to come. But shortly before 6:24 p.m., the subvortex appears before them seemingly out of nowhere, as it curls into its second apex. The tip of the apex brings a 200-mile-per-hour killing wind down upon their location on Reuter Road, near a shallow, cottonwood-lined creek. For the next twenty seconds, with the car now inside, the subvortex remains nearly stationary. Its core lofts the Cobalt and carries it south into a field, then east, and northeast over Reuter. The car hurtles, plunges, and tumbles for some 656 yards, before coming to rest in a canola patch.

CHAPTER TWENTY-FIVE

HOW FAR FROM DAYLIGHT

DAN ROBINSON KNOWS nothing of the headlights behind him, or of their disappearance. There is only the way ahead, the darkness to the south, and the race forward that will deliver him to daylight or the beast in his rearview. The rain falls not in a single, continuous sheet, but in battering waves, and the Yaris shudders as each successive series breaks over the frame with the force of a tide. In these moments, he sees nothing beyond the windshield, not even the hood of his own vehicle. He drives on in blind faith, praying the road is clear.

The Yaris's safety features seem determined to betray him. The traction control system detects the slippage induced by the wind and the mud and reduces engine power to one or both of the front wheels. Robinson backs off the gas to disengage the system and floors it again, repeating this maddening process every few seconds, so that he never attains a speed above forty-two miles per hour. He blows past the stop sign at Alfadale, then, an interminable mile later, the one at Radio.

The Yaris fishtails in the current. The tornado core and the sub-vortex are closing in, and he can see that the rain curtains riding the outer circulation are drawing shut across the road ahead. He hits the gas and the brakes—and hits it again. The wide panorama of the southern plains is lost to a close-in wash of swirling gray and the white spume of aerosolized water, and there is no way to know what wind it carries. "I've gotta get out of whatever this is," he says, punching the traction-control manual-override button. "The car just isn't *going*! Traction control *off*!"

Shortly before 6:24, he breaks through. The rain softens and the wind decelerates. Light creeps into the view behind, and with greater distance the eastern flank of the tornado reveals itself. It is the kind of sight Tim would have exulted in. Crepuscular shafts from the sun sift down onto the fields. A thin ribbon of horizon ends crisply at the southern edge, where the inflow feeds into the circulation like a smoking river. Skyscraper-size crags of dust slide from north to south across miles of tornado flank, drifting loosely, then hardening with the suggestion of vortical motion. Somewhere inside, a subvortex remains nearly motionless at its apex, but Robinson can't see it from here.

He brakes and finally looks at the thing that nearly killed him. *"I just drove through that!"* he shouts.

Sensing an eastward tilt, he drives another few hundred yards down the road before stopping again. He grabs his camera, steps out, and begins filming. As it clears Reuter, subvortices coalesce, one after the next, into bone-colored columns moving at speeds for which there can be no response. The wind picks up again, the trees along the road fold over, and Robinson staggers backward. He sprints toward the ditch and dives to the ground. The rear window of the Yaris explodes under a barrage of hail, and the world around is waylaid with scouring braids of red dust and ballistic ice.

A hailstone opens a freely bleeding gash on his face, but that's the only scratch he'll suffer today. He's made it out.

A few hundred yards at most separate Dan Robinson from the killing wind. To his west, just a few hundred yards separate Tim, Carl, and Paul from daylight.

———

Moments after the subvortex enters Reuter Road, Howie Bluestein and his graduate assistants park RaXPol on the Interstate 40 off-ramp at Banner Road, no more than a few miles east of the circulation. Jeff Snyder sets a zero-degree elevation angle for the scan, so that some of the beam is directed into the ground, while the rest probes as low as the trees permit.

The tornado has transformed dramatically since they turned their backs on it roughly ten minutes ago, at 6:15. At its widest, the span over which wind velocities exceed 110 miles per hour is 3.1 miles in diameter. To the extent that such a tornado may be assigned boundaries, the vortex has peaked at 2.6 miles in width, the largest ever recorded.

If they study the southeastern edge closely, they will detect what is in all likelihood the subvortex responsible for the deaths of Tim, Carl, and Paul. It is all but invisible translating across the face of the parent tornado, except for that moment during which the orbit allows the dim southern sky to reveal its detachment. Though they can't know this in real time, shortly after 6:25 p.m., RaXPol detects wind velocities in the subvortex in excess of 300 miles per hour. Even more remarkable is the rate at which it rotates around the southeastern flank of the parent tornado: 175 miles per hour, the fastest translational velocity on record, and roughly equivalent to the takeoff speed of a Boeing 747.

"That is a *violent* tornado," Snyder observes. "It looks like it's moving due east now. It may be hooking northeast. We might be able to do an intercept on I-40 if it . . ."

He pauses and watches a current of cloud race low over the fields and into the southern flank. The beast is moving north toward the populous suburbs. Snyder's hands are shaking.

The tornado then undergoes another structural change. The subvortex becomes difficult to track at 6:26. In its place, a ring of smaller suction vortices—as many as five or more at a time—are detected by the radar as scallop-shaped debris signatures. These are untraceable by any but the fastest-scanning mobile radar. RaXPol produces an image once every two seconds, and even then the suction vortices are spectral presences. "You're lucky just to track one vortex for two, three, four frames," Bluestein says. "They're moving 150 miles per hour, and each lasts only three or four seconds. Do the math. It doesn't take long for one to go all the way around the tornado, and they don't even go all the way around. They tend to start in one place, rotate partially around the tornado, and disappear, and new ones start. It's like a merry-go-round."

Inside the suction vortices, Bluestein will later discover, are wind velocities on par with the highest ever observed. After two minutes of scanning, the radar indicates it is time to move. They retract the hydraulic outrigging and prepare RaXPol for the drive. The leading edge of the tornado is within a mile, moving rapidly northeastward.

Snyder sees a wall of rain, the dust at the surface whipping like breaking waves. They pull back onto an emptied Interstate 40 and push east, the antenna still revolving, still collecting the signal.

Leaves and small pieces of paper and housing insulation begin to float down onto the road ahead. "Don't worry about the speed limit," Bluestein instructs. "Go as fast as you can."

————

By 6:43 p.m., the tornado is no longer detectable by radar; the Oklahoma City metro area averts the unthinkable—winds exceeding 300 miles per hour ripping across congested highways. By the time it reaches Interstate 40, the storm has grown so massive, has spawned so many new storms, that a torrent of cold outflow finally drowns the mesocyclone.

The damage path stretches largely through unpopulated areas of Canadian County. Eight are dead and a dozen more have drowned in the resulting flood. Included in the body count are the first three storm chasers ever to die in a tornado.

CHAPTER TWENTY-SIX

GROUND TRUTH

C ANADIAN COUNTY SHERIFF'S deputy Doug Gerten locates the
Chevy Cobalt in a field south of El Reno shortly before 7:00 that
evening. He's sitting in his Crown Victoria, a stout, seasoned investigator wearing a ball cap over his buzz cut. He watches marble-size hailstones shatter like glass against the hood as what's left of the storm
moves away to the northeast, toward Interstate 35 and points just west
of Oklahoma City.

Some forty yards to his north, Tim's sedan is crushed nearly beyond
recognition. Pieces are scattered across fields of wheat and canola on
both sides of Reuter Road. For fifteen minutes Gerten waits, until finally the hail stops falling. Then he steps out onto a prairie coated with
ice in summer and wades through canola that has been matted flush
with the earth, stumbling in its viney tangles.

He approaches the sedan and notes that the front end has been
ripped away. The motor is gone. The rear end and the trunk are either
in another field or have been compacted and thrust into the cabin.
Little more than the chassis remains, and all but one of the tires is missing. The roof of the sedan is now no higher than Gerten's hip. *It took the
car apart, except for the stuff that was welded together,* he thinks.

Through the rear window, he sees a body on the front passenger side. He circles around and peers inside at a man lying prone on the seat, which has snapped and now reclines into the back. His legs are folded in the floorboard, draped in a deflated air bag. He looks middle-aged, with short gray hair that has gone white at the temples. He wears no shirt and no shoes, but his seat belt is still on. Gerten doesn't need to feel for a pulse because he knows what death looks like. He notices broken limbs and lacerations, but no other terribly significant injuries.

He calls the sedan's tag number in to the dispatcher to see if he can determine the man's identity. The registration comes back to a woman from Colorado, named Kathy. He thinks he knows the name Samaras, though he isn't sure how.

Gerten spots a Jotto Desk affixed to the center console, which signals storm chaser; he knows they often mount laptops to such desks to track forecasts. He reaches inside the car and pulls a wallet from the man's back pocket to find his driver's license. He looks from the license to the man's face, visible only in profile. Gerten knows who he is now. On that show about storm chasers, he had often seen this view of Tim's profile when he stared up at tornadoes.

From this moment on, when Gerten communicates with dispatch about Timothy Michael Samaras, he'll use his cell phone instead of the radio. Chasers often carry police scanners, and Gerten worries that if they hear about this, they'll converge on the location. He dials dispatch and calls in the "signal 30."

All through the evening and into the wee hours, Gerten remains with Tim, waiting on the state medical examiner's office. A wrecker should be coming by to pull the Cobalt from the field. The fire department is on the way; they'll free Tim from the sedan's mangled chassis with the Jaws of Life. After an hour or two on the scene, Gerten gets a call from a fellow deputy up the road. The man has found Carl. His body is roughly 500 yards west of Tim's, half-submerged in a ditch running high with rainwater. Paul won't be found until the next morning, once the water has receded.

The night deepens as the firemen come and go, lighting up this stretch of dirt road in the farm country south of El Reno. In a few short hours, the sun will rise on a bad landscape that's only going to look worse in the light of morning. Gerten stays with Tim through it all.

As he waits by the Cobalt in the dark, he remembers for some reason the opening credits of *Storm Chasers*. This is the memory he wants to linger after Tim Samaras's last ride. The guys are standing together on an empty plains back road a lot like this one, posing in front of Tim's gleaming GMC diesel, arms crossed and brows furrowed. There they stand in Gerten's mind, looking to the horizon, raring to light out toward the next target somewhere off in Kansas.

———

The next day, at around one in the afternoon, a young weather service researcher named Gabe Garfield comes across the wrecked sedan as he's conducting a damage survey. He snaps a picture and sends it back to the National Weather Center. The image will soon spread throughout the meteorological community.

By nightfall, rumors have begun to swirl. The scuttlebutt holds that not only have three chasers perished at El Reno, but they were three of the most well-known and respected of them all. Marc Austin, at the Norman forecast office, sees Garfield's photo and shares it with his wife, who soon sends it on to Ed Grubb. Ed had driven a mesonet for years. "The wheels, the door handles, the color . . . it all matches," he tells her.

Too restless to sleep, Garfield and the Austins meet at an IHOP for coffee and try to convince one another that they are wrong. But early Sunday morning, June 2, as the sun rises, Tim's brother Jim confirms the chaser community's worst fears with a message on Tim's Facebook page.

By noon, fellow chasers are combing the muddy fields north and south of Reuter Road, sweating in the oppressive heat and humidity. Marc and Sharon Austin are there, as is Garfield, their friend Erik

Fox, and as many as a dozen other volunteers. The three probes are quickly recovered near a shallow creek. Later, another storm chaser finds Carl's camera, lodged in the littoral mud.

The day after, Kathy Samaras and her daughter Amy arrive in Oklahoma to collect Tim and Paul. Jim has already arranged for a mortuary to prepare them for transport. That night, the Austins host Kathy and Amy at their home in Norman. They take them to dinner and are joined by Garfield and Fox, who's a former cop and veteran with past training in investigations like the one at hand. Kathy has many questions, but they are reducible to one: How could this have happened to a chaser as skilled and experienced as her husband? Garfield does his best to explain. He had assumed someone else would take on the investigation, but he knows now that he and Fox have more access to the evidence than almost anyone.

Tuesday morning, Garfield and Fox return to Reuter Road. Fox approaches the still-saturated fields as though they are the scene of a terrible vehicle accident. He surmises the sedan had been carried south from the road and dropped trunk-first near a bend in the creek, where Paul, the probes, and most of their belongings were found. As Fox walks east, he locates six impact craters, each an average of 150 feet apart and gradually hooking to the northeast. The divots in the soil seem to follow the orientation of wheat lying flush in the direction of the wind. They lead across Reuter and north into the canola field where the sedan was found. It is obvious to Fox that the car hadn't been airborne the entire 600-meter distance. This looks like a high-speed rollover, only worse than any he can imagine.

Meanwhile, Garfield gathers evidence to track the Cobalt's movements that day. The video clips from Carl's camera prove invaluable, while Dan Robinson's rear-dash-camera footage allows Garfield to anchor their locations in time, and to orient them to the tornado center. He drives the route they took, turn for turn, and imagines he were in the car with them. Slowly, the portrait of a storm chase emerges, and Garfield begins to comprehend a series of decisions, each representing a link in the chain that would finally bind them. The rain wrapping around the northern flank ensured they were blind to the evolution of the monster to the south. The speed and erratic movement of the storm shrank the margin of error to near zero.

The dead-ending of Reuter at the airport was a critical blow, forcing them to lose precious ground to the tornado. It meant that Tim would never get ahead of a storm that was gaining speed. "You go south one mile, you go north one mile, and that's two miles of driving time you lost," Garfield says. "That corresponds to two and a half minutes." At the end, those minutes meant everything.

Tim and Carl had a chance to escape, either by bailing early up Reformatory Road or by turning north onto US 81—though either would have meant aborting the deployment or attempting the intercept by taking a far more circuitous approach through rush-hour traffic.

As the Cobalt approached Route 81, the tornado was turning sharply northeast, then almost due north. Any chasers to the south of the storm would have known exactly what was happening. The rear-flank downdraft surge came screaming around the south of the me-socyclone, carrying with it an advancing wall of dust. Bruce Lee and Cathy Finley have dedicated their careers to this storm-scale feature, the one Tim feared would drive the vortex into a rope-out to the north. By US 81, Tim may have thought he'd already seen the storm's shift north, but Lee and Finely believe the RFD gave it second life. It was the accelerant poured onto the flame. The tornado swelled to more than two and a half miles in width, and its forward speed doubled. The sub-vortex began its exterior orbit. It reminds Lee and Finley of Quinter, Kansas, the only tornado that has ever truly frightened them. "We'd never seen anything like that before and haven't since," Finley says, "Except for El Reno." The only difference is that in Quinter, the turn to the north had saved their lives.

After the surge, Tim, Carl, and Paul's last ride turned into a race they couldn't win. Beyond US 81, they had one last chance to bail north to safety, along Alfadale Road. But here again they continued eastward, either because they didn't understand how dangerous the intercept attempt had become or, if they did, because it was too late to make the turn. "By the time they figured out what was going on, they had ten seconds before they got to Alfadale," Garfield says. "That isn't a lot of time to brake and go north if you're driving fifty miles per hour."

Had they stopped shortly *after* Alfadale, they would have endured hellish winds, but they might have avoided the subvortex and survived. Yet they kept going because they had always been able to outrun tornadoes in the past. They had always been able to wriggle free. "They put their noses in some places we just wouldn't," Bruce Lee says. "Obviously, that comes with a price."

By the time they were halfway between Alfadale and Radio, the fates of Tim, Carl, and Paul were fixed. "Storm chasers all think alike. It's clear to me why Samaras did what he did. He wasn't being irresponsible. He had to walk the edge. He'd walked it a hundred times before and came out just fine," Garfield says. "If there had been an east option where the airport was, he'd have been just fine. If the tornado hadn't turned north, they'd be fine. If rain hadn't formed on the north side of the tornado, and if they hadn't had their probes that day, they would have been fine. Because all of those elements came together, it all stacked up against them."

But this sequence doesn't even begin to describe how vanishingly remote the odds of what happened to them really were. For scale, the parent tornado was 2.6 miles across, which is wider than Manhattan. The subvortex, meanwhile, was roughly the width of a couple of football fields. Yet in all that survivable space within the parent tornado— scary though it would have been—they encountered the subvortex at one of only two locations where it would have been nearly stationary, and where they would have been exposed to its lethal winds for a protracted period.

Whether you believe in fate, bad luck, or neither, this was an unlikely outcome. It's a little reminiscent of Jarrell, Texas, and the Double Creek Estates. The neighborhood was so small, and there was so much open farm- and ranchland surrounding it. Yet the tornado found that little postage stamp of a subdivision. So it was with Tim, Carl, and Paul. They ended up on that particular stretch of Reuter Road, the worst place in the storm, at precisely the wrong moment.

Once the car was overtaken, the details are grounded in fact but open to interpretation. Kathy Samaras asks Tim Marshall, one of Josh Wurman's collaborators, to glean whatever information possible from the vehicle's event data recorder, commonly known as the black box. She wants to know if there was something wrong with the Cobalt. Upon examination, however, the recorder gives no indication of a flat tire or any other mechanical malfunction that would have prevented escape. The transmission, though, holds one clue: it was in reverse, and the black box seems to indicate this gear shift happened shortly before the recording ended. "My guess is they probably knew they were in trouble," Kathy says. "I think they saw something that made them realize they couldn't go into what they were going into, and they had to get out. 'We have to put it in reverse and get out of here.'"

A second theory comports with a last-ditch strategy employed by chasers who know they're going to take a hit. In 2004, Tim was in pursuit of a tornado near Crystal Springs, Kansas, when he was overtaken by a rain curtain. He shouted to the driver, "Turn your car into the wind." He wanted to give it the profile of least resistance, rather than allowing the broad, aerodynamically clunky flank of the minivan to take the brunt of the gale.

"The winds are coming out of the north, and Tim Samaras knew that," Garfield says. Perhaps as Carl was backing up to angle the Cobalt into the wind, the subvortex came unexpectedly out of the east and caught them broadside.

Of course, there is a simpler third scenario: that the reverse gear was accidental, caused by tumbling metal against earth.

One of the few facts known with any certainty is that the three never had a chance to deploy the probes. The devices were all switched off.

More than a month later, Doug Gerten, the sheriff's deputy who found Tim's body, takes another drive along Reuter Road. He stops near the creek and walks along its banks. Pieces of the Cobalt—a car-stereo speaker, a headlamp, a bumper—still litter the fields and the ditch. Down in the creek, he spots something black partially submerged in a few inches of water. From the shallow water, he plucks out Paul's camera.

What he holds in his hand is the final record. There is one thing about Tim's son that everyone knows: he was always filming and probably had been up to the last moments of his life. These may be stored on the camera's data disk, or they may have been erased by weeks spent in the water. "I talked with Kathy about that," Garfield says. "She does not want to know what happened. She's content with what she knows."

————

The deaths of Tim and Paul are a hard blow to everyone in their family and the chasing community. For Tim's son Matt Winter, there is a sense of whiplash. It had been only seven years since he'd met his biological father. They'd had so many years to catch up on, and years to do so— but now that time is gone. Tim and Winter hadn't talked much during 2013. On May 20, eleven days before Tim died, Winter had sent him a picture of his five-year-old son, Peyton, climbing an archway in the house like a little monkey. Winter thought it was funny because his older son, Nick, had done the same thing at Tim's Lakewood house the first time Winter brought his family to visit.

"Does this look familiar?" Winter wrote.

"Yes, very much so," Tim had responded. "Looks like a certain other little guy I know."

On May 31, Winter remembered pulling up the radar feed depicting the storm over El Reno while he was at work at Nationwide Insurance. To him, it looked like a hurricane—bigger than any other mesocyclone he had ever seen. A couple of days later, Jenny called. He took the call in the conference room, and she tried to tell him the news—that his father and half brother were gone—but she couldn't get out the words. Kathy picked up the phone and finished for her.

Winter sagged into the nearest chair, and a wave of nausea washed over him. When he hung up, he told his boss he had to leave and walked outside to the parking garage. It was a sunny day, but he says he heard the concussion of thunder. He ran to the edge of the parking garage and looked up. "It's one tiny, low-topped thunderstorm over Des Moines," he says. "It was so random. Dew points were way too low to support any kind of development. I felt like it was some sort of sign, maybe from him, saying good-bye."

But by the time he arrived home, the sign had faded, and no explanation seemed enough to encompass what had happened. When he sat down with his children, they just couldn't understand. They asked the questions he didn't know how to answer:

"How could that happen to him?"

"Wasn't he the best?"

"He was with Carl. Isn't he good, too?"

"Yes" was all Winter could say. In his heart, he would always feel the same way.

CHAPTER TWENTY-SEVEN

THE SIGNS

E D GRUBB KNEELS on the front lawn of Kathy Samaras's Lakewood bungalow and dismantles a mesonet station in the light of
a late-winter afternoon. The grass is thatched with shade from the tall
maples, and Kathy sits nearby in a wooden rocker with the sun on her
face. On this day in February 2015, she wears short sleeves, though
snow still clings to the shadowed eaves of a house across the street. The
neighborhood is settled deeply into the weekday quiet, the neighbors
at work, their kids at school.

Kathy, Ben McMillan, and I watch Grubb break the mesonet down
into its constituent parts: anemometer, barometer, and the PVC pipe
that inducts air through the temperature and relative-humidity gauges.
Bruce Lee and Cathy Finley will receive the parts in Minnesota. It is
unclear when or if TWISTEX will ever embark on another mission.
Tim was the research entity's sole fund-raiser, and no one so far has
stepped into his role. Even if Lee and Finley secure a grant, the mission's shape may be fundamentally different. Kathy retains possession
of Tim's probes, and she is deeply conflicted about having anyone else
attempt the very thing that took the lives of her husband and son.

When Grubb's task is finished and the parts are placed into the back of a friend's SUV, Kathy goes inside and sits at her kitchen table in front of Paul's Mac computer and a glass of iced tea. Her face is reflected in the darkened screen for a moment until she boots up the computer and begins to sift through its contents. The last person to have done so would have been Paul.

One of the first photographs she finds was taken on May 9, 2013, a few weeks before Tim and Paul were killed. In the frame, a single oil-well pumpjack attends a shaft of lightning stitching through the underside of a supercell down into some unheralded corner of the plains. There is a white Chevy Cobalt, one of the mesonets, parked in the foreground. Paul was shooting a long exposure, and Tim must have gotten out of the vehicle halfway through. He stands next to the Cobalt, looking up into the hardening clouds, his body translucent, spectral.

There's a self-portrait of Paul, too. His black beard is thick and lustrous. Kathy wishes she had more pictures of him. The problem is, he was always the one behind the camera.

She opens the next file and finds clips from an unfinished documentary Paul was filming about his father. In one segment, Tim is in his shop, working on the clear dome turret for the Lightning Intercept Vehicle. He's wearing faded jeans with gaping rips in both knees, and a denim shirt with National Technical Systems' logo embroidered on the chest. His glasses tend to slip down his nose, and he pushes them back into place. He's fifty-five years old here and still trim, but his cholesterol is a little too high. The hair at his temples has gone white. He wears it shaved close at the sides and a little longer on top, which gives him a distinguished, even professorial, bearing.

Behind the lens, Paul's camerawork is graceful and effortless. The frame glides over the workbench and dives in for close shots of Tim's hands. His fingers are thick, his nails dirty, like a mechanic's. Big veins shunt across his thickly muscled forearms.

In another clip, Tim is holding his grandson Jayden, at his first birthday party, and smiles contentedly. Jayden reaches for Paul and the camera lens. "You're putting fingerprints on the camera," Tim lovingly scolds, then brings Jayden in closer, and the child laughs, his tiny hands outstretched.

One of the last videos Kathy plays is of Tim and Carl as they prepare for their final season. On May 17, Ed Grubb is squatting atop the probe truck, working on a mesonet station in the driveway of their house in Bennett. Tim is showing Carl around his shop and the recent improvements he has made to the vacuum system. He leads Carl to the truck and bemoans a side panel's leaking compartment. In 2012, Tim had been forced to put their suitcases in garbage bags to keep them dry. But he is evidently excited about the high-speed camera he plans to mount to the dash for the PhOCAL project. "I want to put my Phantom camera in here—when we're not chasing a tornado—if a late-season MCS [mesoscale convective system] drifts over central Kansas," he says. "We'll get a lightning hit on a wind turbine. That's what they really want."

At one point during the tour, Carl asks Tim whether he and his family will remain in this sprawling estate in the foothills. So far from town, the house can feel empty, especially with the girls moved out, and Tim and Paul so often on the road. Kathy has made clear that she doesn't have an attachment to the land the way he does. "Uncertain," Tim replies in the video. "The boss doesn't like it out here. I love my wife more than I love this house, and if my wife doesn't want to live here, that's all there is to it."

At this, Kathy brings her hands to her mouth and gasps. Grubb reaches out and takes her in his arms.

———

In the heat of midsummer, I find Kathy leaning on a shovel in the front yard in Bennett. She has just buried a field mouse whose body she'd found in the house that morning, and she is taking a moment to enjoy the solitude. "It's so quiet out here," she says. The sound of the wind over the high foothills east of Denver takes on a deep resonance.

She leads me into the four-car garage where Tim spent most of his time listening to Bruce Springsteen and Led Zeppelin on satellite radio. "He'd get up and he could spend all day here," she says. Nearly every square inch is devoted to his tools: band saw, drill press, lathe, Miller welder. The tubes inducting sawdust and metal shavings into his custom vacuum system crawl along the walls. Sunlight pours in through the garage-door windows. His workstation looks untouched, its Panasonic Toughbook still plugged in.

Tim's shop was only supposed to fill one side of the garage—there's a wall dividing it down the middle—but it had spilled liberally over to the other half. A mesonet rack is propped against a wall. A white Chevy Cobalt—Kathy thinks it's M3—occupies one of the spaces.

We exit the garage through a back door into the house. At the threshold are Tim's work boots and Carl's Vans. The place looks as if they might return from a chase at any moment, and their things are both a comfort and a torment. Kathy has long since moved back to the bungalow in Lakewood. This place is too quiet now; it feels like the relic of a part of her life that is gone. Yet, giving anything away is as hard as the keeping. "It seems silly," she says, "but it's like giving part of Tim away. He used it and I can envision him working on it." Somehow, she will have to clear everything out and ready the house for sale. Tim was a pack rat, so this is no mean task. It sometimes seems so enormous a job that it's paralyzing.

The office is the way he had left it: orange HITPR on the floor, a panel of computer screens, and a massive hard drive whirring softly. A calendar on the wall is filled with appointments: a National Geographic talk in Maryland, the week after El Reno, and another in Chicago, October 7–8; a ham radio convention in Estes Park, June 29. The months during tornado season are devoid of appointments. His goals are listed on a dry-erase board:

Build camera platform.

Order five ball mounts.

We wander outside, where the probe truck is parked in the driveway. It is an impressive machine: powerful diesel engine, LEDs lining the windshield, spotlights mounted to the grill guard, and Tim's good-luck charm, a McDonald's cheeseburger, sitting on the dash, as stiff as a hockey puck.

The sight of the truck gets Kathy thinking. "I knew they were taking the car. I didn't see what was in the car," she says. "I didn't think to ask why because they'd taken the cars before. That was my question to Gabe: 'What if they'd been in the truck?'"

The answer didn't prove as simple as the question, though. The diesel engine might well have allowed them to fight their way to daylight; or the GMC's high profile might have presented the wind with an even bigger target to batter around.

There are a lot of what-ifs: about the car, the roads, the storm, the decision making; about the day of and the years prior. What if any one piece hadn't been arranged just so?

Then there are the hints only obvious in hindsight. Over the years, Kathy saw more than one video clip that concerned her: "He'd come home and he'd have video and he'd be showing it," she says. "There were times it seemed he was too close." She would tell him as much, "and he'd go, 'Oh, no, we were so many feet away.'" He'd tell her all the reasons it had been safe. She wasn't sure she agreed with him—and it looks so clear now, staring back at it.

But why not trust him in those moments? He was an expert, so were his companions. And in all those years before 2013, no chaser *had* been lost in a tornado. Even in all the conversations they had about safety when Paul joined the crew, "We never ever discussed *this* ever happening," Kathy says.

They'd had plans to travel, Kathy and Tim, to see the country together, to spoil their grandchildren—maybe even to see Paul settle down with the right woman. Tim had been talking about spending less time on the road in the coming years. He was getting older, and the field work was getting harder and harder on his body. What if he'd chosen to dial it back a few years earlier? Would they be on some beach together right now, far from the landlocked plains and its howling winds?

There's a temptation to pick through each what-if. There's a temptation for the mind to travel down the counterfactual path—be it out of longing or doubt or anger—to find a route that leads away from that fateful stretch of road, toward a world with Tim, Paul, and Carl still in it.

But Kathy recognizes the futility of these alternate realities.

"It doesn't matter," she says, "if anything could have or would have been changed." The outcome is the outcome. Wherever Tim and Paul are now, Kathy knows, that's where they'll be. No matter how you turn over the past.

It holds true even for the mother of all what-ifs—that tortuous question that everyone in Tim's circle brushes up against at some point: *What if I could have done something to stop him?* It's this hypothetical, in a sense, that underscores the wisdom of Kathy's approach and the trouble of searching for the pivotal what-if.

Tim says on video that he would have moved back to Lakewood for Kathy. If she had asked, he would have agreed in a second. But what if she *had* gone beyond, and asked the harder question of him?

Imagine that she had known everything in advance: the risks he was taking; the dangers that even he could not see; the ultimate outcome toward which he was hurtling. Imagine it was all there to understand and reckon with, one year before or ten. Even knowing all, would she—or any friend or relative—have been able to halt Tim's momentum, to convince him to turn back?

Would anyone, the world round, have been able to make Tim stop chasing?

Even the man, himself?

———

Kathy and I go inside. We walk upstairs and out through the master bedroom's French doors, onto the balcony where Tim once said he could see storms clear to Kansas. A PhOCAL sprite camera is angled off to the east. We can see the snowcapped Front Range. A mourning dove pipes dolefully. Kathy looks out over the rearing foothills. "This, I'm going to miss. On nice days you can see the wind turbines," she says. "This was his dream. It's where he wanted to retire. But every time I get out here, I get sad. He's not here. Paul's gone."

Kathy thinks about whether she will see her son and husband again. She believes in a hereafter, where they'll be reunited someday. Tim had been doubtful about the prospect. "He thought, 'If you're dead, you're dead, and if you live, you live,'" Kathy says. Whether or not there is an afterlife, he did have a very specific, very *Tim* request for the handling of his remains.

"The one thing he used to say is, 'If I die before you do, take my ashes and have somebody put them up in a tornado.' That's what he used to tell me." She pauses for a moment, as if she's imagining what he might say now, after all they've been through.

"Well, he's already been in the tornado, so I get to do whatever I want with his ashes."

Twice now, Kathy has felt his presence, some signal that he's been watching over her. The first was when Ben McMillan, Ed Grubb, and Tony Laubach sent her roses. Due to some delivery snafu, they arrived on their wedding anniversary. Tim had always brought her red roses on their anniversary.

The second time was last night. She was dreaming. Kathy opened the front door and Tim was standing there, looking as though he'd just returned from a chase, waiting to be let in.

CHAPTER TWENTY-EIGHT

TIM'S LEGACY

THE CORE OF a violent tornado was once a mystery to us, as un-reachable as the surface of the sun. For decades, some of the scientific world's brightest minds struck out for the plains like hunt-ers. Their weapons of choice were Rube Goldberg contraptions in all shapes, materials, and sizes, designed for a single purpose: to extract knowledge from one of the harshest environments on the planet. But the tornado's mysteries proved stubbornly elusive. Back in the early eighties, the barrel-shaped TOTO toppled in a weakly tornadic wind. "There are easier ways to do this," Howie Bluestein had lamented. A second effort in the nineties, part of a multimillion-dollar, federally funded research expedition, used an aluminum-alloy tube mounted to a steel plate and caught a glancing sample at the edge of an F4. Oth-ers tried, each with their own particular gizmo, and each failed. The hunters moved on. Their quarry was too unpredictable, too danger-ous. They sought other tools, even though they knew all too well there could be no substitute for probing the core directly. The only way to understand what happens in that place where houses fall and people die was to get *inside* it—then to emerge again, all the wiser. But maybe such a thing couldn't be done.

No one—aside from his fellow chasers—had ever heard of Tim Samaras when he appeared on the scene with his "turtle." Yet this nobody—a man whose academic pedigree began and ended with a diploma from Alameda High School—pulled off one of the most dramatic coups atmospheric science had ever seen. The core was not, as he proved, untouchable.

He spent the remainder of his life building on his breakthrough at Manchester, and pushing the in situ probe field forward. He succeeded with a frequency that no one has matched since. "What Tim was able to do," says Tim's TWISTEX partner Bruce Lee, "blows everybody else away."

Lee once said that knowledge isn't advanced by one spectacular measurement, or even two or three. Solving the mysteries of tornado formation, intensification, and decay is incremental work. It takes dogged persistence to track down and sample a menagerie of vortices, from the single-cell, to the two-cell, to the multiple-vortex, and every subspecies in between. One of the many tragedies of Tim's loss is that he died before he could finish this work, to the extent that the labor of science is ever complete. Near the end of his life, he was developing a new anemometer design, one without moving parts, built to withstand high-end tornadic winds. But the design, in his exacting eyes, still required a few tweaks. He never got the chance to test it in a twister. In the late spring of 2013, before the anemometer was ready, the storms called him away.

One couldn't be faulted for asking whether Tim failed, or whether his death stands as final proof that probe work is too dangerous, too impossible, to undertake.

The scientific community, however, has already answered these questions since his death. If anything, Tim Samaras reignited a whole new mode of study. Gabe Garfield puts it this way: "We weren't seeing what was going on inside of the tornado where it matters most. Tim showed that it could be done. He laid an important cornerstone at the foundation, and others will build on that." His work redefined what was possible for tornado researchers.

If there's any one person who's been the most diligent in building on Tim's work, it's none other than Josh Wurman. As often as he and Tim failed to see eye to eye, Wurman has continued down a path Tim blazed. The founder of the Center for Severe Weather Research has kept expanding his own fleet of pods, deploying them into the paths of tornadoes while simultaneously scanning with his Doppler on Wheels. Mobile radar provides the safety net for Wurman's pod team that Tim lacked at El Reno—a precaution that Wurman and other researchers have come to see as a necessary foil to the hazards of probe work.

That Tim was willing to operate without a safety net for so long is an enigma to some. Even more vexing is that he saw his fate coming, at least on some level: "Somebody's gonna get bit," he had said, presciently, even as he pushed his limits further. The contradiction perhaps speaks to the double-edged nature of obsession, and how it both drove and doomed Tim. His devotion made possible unheard-of feats of scientific discovery, but it also inured him to the risks he was accruing—as he grew calm and comfortable under the anvil, as his son entered the same perilous territory. The effect of his own expertise was a kind of tunnel vision. Looking out, Tim was keenly aware of the looming threat, but blind to his own proximity.

After so many years of inching closer, he couldn't appreciate how close to the dragon he had finally sidled. He could reach out and feel its presence—the warmth of its breath, the tick of its pulse. He could grab ahold of it, and it of him.

———

On May 9, 2016, Tim Marshall stared down a tornado near Sulphur, Oklahoma, watching for the left or right drift. He dropped his last remaining pod on a country road and fled with the wind at his heels. Based on the surrounding damage and an examination of DOW radar data, the pod caught the outer edge. That day, Marshall claimed one of the closest hits in history—and the first since Tim's death. The scientific effort marched forward, one step more, one step wiser, though perhaps without the same verve. It was no Manchester; it wasn't a core strike.

As difficult, perilous, and often frustrating as this work can be, it's more important now than ever. Recent research into our changing climate, and its effect on severe-storm activity in North America, is sobering yet inconclusive. We know that as the oceans warm, increased rates of evaporation will flood the skies with elevated concentrations of moisture. This means more instability, more CAPE, more fuel for violent storms. But there are also studies that project an attendant decrease in wind shear. Tornadoes require converging air masses to form. Most climate scientists speculate that the result may be a lower overall number of tornadoes.

But there's an important catch: when the elements do align, the tornadoes and the outbreaks that result may be much, much worse. One analysis indicates that this is already happening.

Two statisticians, Elizabeth Mannshardt of North Carolina State University, and Eric Gilleland of the National Center for Atmospheric Research, recently compiled more than forty years' worth of atmospheric soundings, focusing primarily on the two biggest supercell indicators: wind shear and CAPE. When they plotted tornadic storms over these years, what they found was alarming. The "return period," or the average amount of time that elapses between a given extreme tornadic event, appears to be decreasing. In other words, the extremes are becoming less rare, a trend line that may well worsen.

As case studies, Mannshardt and Gilleland examined two recent tornadic events: the May 20, 2013, Moore, Oklahoma, supercell, and the twister that claimed the lives of Tim, Carl, and Paul near El Reno. According to their analysis, these storms are exceedingly rare, situated somewhere near the outer edge of statistical probability. Historically, Moore should only see a tornado of that ferocity once every 400 years. But between 1999 and 2013, the city has been struck by two historic EF5 tornadoes and one EF4.

The El Reno event is even more exceptional. For this storm's atmospheric conditions, the return period is once every *900* years. Tim would never have chosen this fate for himself, his son, and Carl, but he would almost certainly have been thrilled that he was present for a nearly once-in-a-millennium event.

The statistical model on which these figures are based, however, assumes a constant climate over the last forty-two years—which is not the case. When the model allows for a steady increase in certain parameters, such as atmospheric instability produced by warming oceans, the research indicates that such massive storms are likely to recur far more frequently as the climate changes.

There is hope, as research into massive tornadoes continues. One of the most promising new research initiatives involves TWISTEX's own Bruce Lee and Cathy Finley. With TWISTEX at loose ends, they partnered in 2014 with a scientist and computer whiz on a first-of-its-kind project. Leigh Orf, a researcher at the University of Wisconsin at Madison, had just made a major breakthrough, using raw atmospheric data and one of the fastest supercomputers in the world, to simulate a supercell and EF5 tornado from birth to death—simply by feeding it a primed atmospheric sounding and letting a state-of-the-art physics program run its course. The wedge in his uncannily lifelike visualizations bears an unmistakable resemblance to the real May 24, 2011, event on which it is based. Orf has brought Lee and Finley aboard because he now needs field scientists who've witnessed the beast in the flesh; his data is so formidable and in such high resolution that he needs experts who know what to look for, and where.

The three now have on their hands the answer to every question a scientist could possibly ask about a single monster tornado. Represented by the ones and zeros of computer code are all the internal currents that mobile radar and even in situ probes could never see. As of this writing, the trio are still hard at work teasing apart and reverse engineering the sky's most complex riddle. The next superstorm they plan to simulate is the one that killed their friends in El Reno.

EPILOGUE

LORETTA YOST'S SPEECH is as austere as the plains—clipped, a little formal, belonging to another century. She wears a simple cotton dress, a black cardigan, and a black Mennonite prayer cap that covers much of her hair, iron gray with streaks as pure white as the snow outside. Her late husband, Harold, was a bricklayer and had built this house himself. It is small but warm, smells of freshly baked chocolate-chip cookies, and is nestled amid the fields. Through the window she can see the stubble from September's soybean harvest.

Loretta opens her journal and begins to recount the day of June 24, 2003, on her family's twenty-four-acre farm near Manchester, South Dakota. Her grandchildren had been splashing around in a small plastic swimming pool on the west side of their rambling, nearly hundred-year-old home. One of her grandsons, four-year-old Jacob, was reclining in a lounge chair near the pool as though he were at the lapping edge of the ocean, and not on the high South Dakotan plains. The place was surrounded by elm, spruce, and cottonwood trees so thick that the house could scarcely be seen from the dirt road.

At around six thirty that evening, Harold, Loretta, their sons, daughters, and grandchildren all loaded into the pickup and drove over to dinner at the Unruh farm, on the other side of Manchester. They brought their wooden ice-cream maker with them to prepare dessert. By the time they arrived, Mike Unruh's weather radio was wailing. A tornado had been spotted a hundred miles southwest, between Fort Thompson and Woonsocket. "*A long ways away*, we thought. Did Mike turn the weather radio off? I don't remember hearing it anymore that evening," Loretta later wrote in her journal.

Liz Unruh made a pizza, and they followed dinner with ice cream, complete with strawberries and chocolate syrup. It was during dessert that the phone rang. Mike Unruh's brother was on the line. "If you look east," he said, "you'll see a tornado."

They filed out of the house and gathered in the yard. Off in the east, they saw it, "a short, fat, gray tornado. I don't remember being scared," Loretta wrote. "We kind of walked down the road going east, children and all. Above, big clouds were building. We scattered out on the road looking east, feeling quite a bit of excitement. Bill Fox came by in his pickup and said he'd heard there was a storm in Manchester, and said he was going over. 'That's where we live,' I exclaimed."

As Loretta spoke, insulation began to flutter down around them like snow. They walked back to the Unruh house. Harold, Mike Unruh, and Loretta's son Ace prepared to leave and see about the farm. Loretta's daughter Eva called their house, but the answering machine never picked up.

The phone rang again at the Unruhs'. This time, it was Harold's brother on the line. He asked Liz, "Are they there?"

"Yes," Liz said. "They're all here."

"Their place is gone."

Loretta ran out of the house. Harold, Mike, and Ace were about to pull away. She caught up to them and relayed the bitter news.

It was their basement into which Kingsbury County sheriff Charlie Smith had called, "Harold? Harold?" The mud-coated basset hound found wandering around the wood and cinder blocks was their dog, Bailey. It was the destruction of their home that shed turbulence into the pressure profile sampled by Tim Samaras's turtle. The defining moment of Tim's life is theirs, too.

Their new brick home was the last big project of Harold's life; he died in 2008 at the age of seventy. Loretta and her son Lockwood offer to take me on a drive. We pile into their minivan and head into Manchester. The old sign still stands, battered, bowed in at the center, its edges curling around the wooden posts. We step into the kind of bitter prairie cold that makes skin sting and bones ache, and we walk toward the monument to a ghost town. Harold had laid its flagstone foundation, and at the center is a marble plaque with raised bronze lettering that tells the story of Manchester's end: JUST PAST SUPPER-TIME, AT ABOUT 7:30 P.M., TUESDAY, JUNE 24, 2003, THE SKY VIRTUALLY FELL ON MANCHESTER. It goes on to say that though the F4 destroyed the artifacts of the township's 122-year history, the neighbors banded together to help each other just the same. As a coda, the last line is a quote: "THAT'S THE BIGGEST DROP EVER RECORDED," SAID TORNADO RE-SEARCHER TIM SAMARAS, "LIKE STEPPING INTO AN ELEVATOR AND HUR-TLING UP 4,000 FEET IN TEN SECONDS."

We climb back into the minivan, and Lockwood steers over dirt roads hummocked with snowdrifts. A John Deere tractor mechanic, he wears tan leather gloves and a canvas jacket that smells of diesel. He brakes at an empty field near Redstone Creek, and he and Loretta look out over what had once been their home.

"The foundation of the garage is still there," Loretta observes.

"The house was east and south of that slab," Lockwood says. "It's been cleaned up, pushed in. The barn was right there. The hog barn was behind it."

"Every time I come out here in the summertime," Loretta says, "I hear the birds sing, the meadowlarks."

"We had a row of poplars clear to the ditch to the north, and maples to the east."

The weak winter sun hangs low in the sky, a dying coal through the haze. I walk away from the car to a spot on the dirt road near the intersection of 206th Street and 425th Avenue. To the east, in a field of nubbed corn, the slouching hulk of a steel granary remains where it had been deposited nearly twelve years before.

I look to the west and try to imagine a minivan fishtailing down a rain-slicked road, its gravel the consistency of cake batter. Tim is behind the wheel, and he's barreling east with an eye on the tornado to his right, intuiting the location where two trajectories will converge. He picks this patch of dirt road, hurriedly deploys the turtle, and speeds away, perhaps watching in the rearview mirror.

He doesn't know it yet, but even as the Yost farm fails, even as the surrounding soil, steel, and wood lift into the sky, neither the howling gale nor the shower of debris will move Tim's remarkable creation. It catches the storm, and it holds fast.

AUTHOR'S NOTE

I N DOCUMENTING THE life of Tim Samaras, from the first appliances he dismantled on his boyhood bedroom floor to his last moments alive, I relied heavily on a network of dozens of intimates—family and friends, chase buddies and colleagues. Having never met Tim myself, I knew him at the start only as the daring researcher on the Discovery Channel's *Storm Chasers*. Without those close to Tim—who generously opened up their lives and shared hours upon hours of their time—this work of journalism would never have been possible.

Nothing in this account has been fictionalized—no characters, events, or dialogue are composites. Any direct quotations are drawn from recordings of Tim, his writings, news reports, and the accounts of those present for the events in question. Any scenes or glimpses into his interior world spring from his own personal correspondence, his or his colleagues' voluminous chase footage, and interviews with the people who knew him best.

Just as essential to recreating this rarefied subculture were my own experiences beneath the storm. In the reporting of this book, I spent weeks on the road chasing tornadoes with some of Tim's best friends. I knew that to understand him, I'd first have to find the swirling wind myself. Over days, over thousands of miles, over too many busts and near misses, I lived in Tim's world. I felt what must have been the same exhaustion, the same boredom, the same disappointment at each storm that failed to deliver. After three weeks I was almost ready to give up. Then, our luck suddenly changed. The next day I found myself in Nebraska, face to face with two simultaneously occurring EF4 tornadoes. I got it then. I have at least some grasping sense, now, of why Tim went out year after year in search of them. I know how this wonder, this adrenaline, can exert a pull as irresistible as gravity. And I'd be lying if I said I haven't gone chasing since, even though this book is finished. I feel that pull still: when I hear the big black clouds are getting ready to boil, I can't help but wish I were there. More practically, those weeks spent with seasoned veterans like Ed Grubb, Ben McMillan, Tony Laubach, and Dan Robinson gave me insight into how chasers think, how they operate near dangerous storms, and how Tim, in particular, maneuvered for the intercept. Though my own adventures with these men are not recounted in *The Man Who Caught the Storm*, they nonetheless inform every page.

ACKNOWLEDGMENTS

First and foremost, I want to thank Kathy Samaras. Over the more than three years it took to research and write this book, I thought of her daily. The memories she has shared constitute the most closely held truths of her life. Many were painful, but her words opened a window of understanding on the man she loved. If this book manages to capture even a flicker of Tim's fire, it's because of her. In equal measure, Tim's daughters, Amy and Jenny, and his son Matt each illuminated Tim's personality and life as a father. And I'm grateful as well for their precious stories about Paul, a beloved brother and son. Indeed, the generosity of the Samaras clan, including Tim's brother Jim, has made this book possible.

I relied heavily on the kindnesses and forbearance of those who knew and loved Tim Samaras, Paul Samaras, and Carl Young. Bob Young, Carl's father, spent hours reminiscing with me about his son. Ben McMillan, TWISTEX's resident EMT and my hurricane buddy, brought me along for one of my first storm chases; we spent an exhausting and often exhilarating week in pursuit of the swirling wind, trekking from one end of Tornado Alley to the other. With Ed Grubb and Tony Laubach, two stalwart mesonet drivers, I had the incredible experience of witnessing two simultaneously occurring EF4 tornadoes in Nebraska. McMillan, Laubach, and Grubb invited me into their singular world and, in turn, allowed me to better know Tim, Carl, and Paul.

The learning curve of atmospheric science is a precipitous one, to say the least, and as a writer, not a meteorologist, I remain stuck near the bottom. To ensure that this book presented the science accessibly for the layperson and accurately enough for the well-informed, I sought out the help of talented researchers and forecasters, none so often as Gabe Garfield, a research meteorologist at the Cooperative Institute for Mesoscale Meteorological Studies. Gabe patiently fielded what must have seemed an inexhaustible stream of questions, not only about the science but about Tim, Carl, and Paul's final chase. Headquartered at the National Weather Center in Norman, Oklahoma, he took up the inquiry himself in the days, weeks, and months after the El Reno storm. He drove me along the route they took, annotating nearly every moment with their words and details about the tornado's location and evolution. As a cheerful and dependable resource, Gabe was essential to my reporting.

Over these years, I've had the opportunity to get to know a vast network of chasers, researchers, and friends whose lives Tim has touched, and whose recollections helped me to piece together the arc of his own life: Bruce Lee and Cathy Finley, TWISTEX's seasoned field researchers, were kind enough to share their unvarnished accounts of the team's most important intercepts—warts and all—including deployment footage that put me in the passenger seat next to them. Along with Tim, they were TWISTEX's heart; it would be impossible to overstate the importance of their testimony. Larry Brown, Tim's longtime boss at ARA, DRI, and NTS, led me through the development of Tim's unusual skillset. Tim's earliest comrades-in-arms, Pat Porter and Brad Carter, told me stories about his early chases, in a time before the Samaras name rang out in the world of atmospheric science. Anton Seimon, a fellow chaser and researcher, was a tremendous resource in more ways than one. Not only was he present for one of HITPR's first missions, his correspondence with Tim yielded essential insight into the mind of a man I could never meet. As if this weren't enough, his incredible website, the "El Reno TED: Tornado Environment Display," acted as a surrogate for my own eyes when I couldn't experience what Tim, Carl, and Paul saw at El Reno. Bill Gallus, an Iowa State professor and one of Tim's early supporters, illuminated for me a milieu where a man with Tim's credentials—or lack thereof—was never entirely at home. Joshua Wurman, whose Doppler on Wheels is the only reason we know exactly what Tim, Carl, and Paul encountered on Reuter Road, contributed a far more nuanced understanding of this book's subject than would otherwise be possible.

Beyond these pages, I'd be remiss not to mention my own support system. My agent, David Patterson, saw from the beginning that Tim's life and works cried out for a detailed account. My editor, Jonathan Cox, is the best I've ever partnered with. Before he came aboard, I was lost belowdecks; he helped me see the rest of the ship. I can't begin to imagine what this book would look like without Jon.

As we wrestled with the telling of this story, I leaned on my wife, Renee, for support, both emotionally and financially; at the end of a tough day, when it seemed like things weren't going so well, she was my shelter. My mom and dad, Laurie and Hal, and my sister, Holly, have always been there for me, and these past few years have been no different. Last, but certainly not least, I must thank my friends and fellow writers, Tara Nieuwesteeg, George Getschow, and Mike Mooney. Sometimes they were the best sounding boards I could ask for. Their feedback was spot-on when I needed it most. I look forward to returning the favor.

NOTES

Prologue

1   *The fire department's siren sounded:* Phan T. Long and Emil Simiu, "The Fujita Intensity Scale: A Critique Based on Observations of the Jarrell Tornado of May 27, 1997," National Institute of Standards and Technology, July 1998, 4.

1   *a shrill, oscillating note:* Interview with a longtime member of the Jarrell Volunteer Fire Department.

1   *The siren was only ever used to call up volunteers:* "Tornado Disaster—Texas, May 1997," *Morbidity and Mortality Weekly Report,* Centers for Disease Control and Prevention, November 14, 1997.

1   *For a time it seemed to track neither north nor south:* Scott Guest chaser video, https://www.youtube.com/watch?v=dfCXofp7Pgw.

2   *The tornado was as wide as thirteen football fields:* Brian E. Peters, "Aerial Damage Survey of the Central Texas Tornadoes of May 27, 1997," C-2, https://www.weather.gov/media/publications/assessments/jarrell.pdf.

2   *School had let out for summer:* Allen R. Myerson, "Town Is Upended by Tornadoes Twice in Eight Years," *New York Times,* May 29, 1997, http://www.nytimes.com/1997/05/29/us/town-is-upended-by-tornadoes-twice-in-eight-years.html.

2   *Double Creek Estates:* Long and Simiu, "Fujita Intensity Scale," 12.

2   *With no choice but to shelter aboveground:* "Tornado Disaster," *Morbidity and Mortality Weekly Report.*

2   *All else was fatality:* James H. Henderson, "Service Assessment: The Central Texas Tornadoes of May, 27, 1997," National Oceanic and Atmospheric Administration, April 1998, 3.

2   *The Hernandez family was the outlier:* Jesse Katz, "A Neighborhood Blown to Nothingness," *Los Angeles Times*, May 29, 1997, http://articles.latimes.com/1997-05-29/news/mn-63711_1_entire-neighborhood.

2   *Their home and some thirty others:* "Tornado Disaster," *Morbidity and Mortality Weekly Report*.

2   *the foundations had been scraped clean:* Long and Simiu, "Fujita Intensity Scale," 11–15.

3   *The carcasses of hundreds of cattle:* "The Jarrell/Cedar Park and Pedernales Valley Tornadoes, Summary of Weather Event of May 27, 1997," National Weather Service.

3   *More than five hundred feet of asphalt had been peeled:* Peters, "Aerial Damage Survey," C-3.

3   *I learned later that the tornado had crawled:* Long and Simiu. "Fujita Intensity Scale," 6.

4   *On average, tornadoes will claim eighty lives annually:* "Tornadoes: A Rising Risk?," Lloyd's of London, February 2013, 19, 21.

4   *the damage caused by tornadoes has outstripped that from:* Rawle O. King, "Financing Natural Catastrophe Exposure: Issues and Options for Improving Risk Transfer Markets," Congressional Research Service, August 2013, 8.

4   *seventy percent of tornado fatalities are attributable to the deadliest breed:* "Thunderstorms, Tornadoes, Lightning: Nature's Most Violent Storms," National Oceanic and Atmospheric Administration, 4, .

4   *An EF5 flattened a swath of Joplin, Missouri:* "Joplin, Missouri, Tornado—May 22, 2011," NWS Central Region Service Assessment, National Weather Service, July 2011, 1.

5   *In Joplin, damage surveyors found a truck:* Timothy P. Marshall, "Damage Survey of the Joplin Tornado: 22 May 2011" (conference paper, 26th Annual Conference on Severe Local Storms, American Meteorological Society, 2012), 15–16.

5   *A Ford Explorer was lofted:* Eugene W. McCaul et al., "Extreme Damage Incidents in the 27 April 2011 Tornado Superoutbreak" (conference paper, 26th Annual Conference on Severe Local Storms, American Meteorological Society, 2012), 2, 10.

5   *It passed clean through the first:* Interviews with Larry Tanner, National Wind Institute, Texas Tech University.

5   *Its frail glass container hadn't even cracked:* Staff Reports, "Tornadoes

Spawn Strange Tales, Some of Them True," *Chattanooga Times Free Press*, April 26, 2012.

5   *In Joplin, a child's play set:* Marshall, "Damage Survey," 16.

6   *false-alarm rate of roughly 70 percent:* J. Brotzge and S. Erickson. "A 5-Yr Climatology of Tornado False Alarms," *Weather and Forecasting*, August 2011.

6   *The people of Joplin had seventeen minutes:* "Joplin, Missouri, Tornado," NWS Central Region Service Assessment, 23.

6   *Xenia, Ohio, received no warning at all:* J. Brotzge and S. Erickson, "Tornadoes without NWS Warning," *Weather and Forecasting* 25 (February 2010): 161.

6   *it had knocked out power to four of Xenia's five sirens:* Staff, "In Xenia, Warning Bells Silenced," CBS News, September 21, 2000.

6   *In 2016, the Storm Prediction Center:* Jeff Frame, "This Is How the 'Surprise' Indiana and Ohio Outbreak of August 24, 2016, Happened," U.S. Tornadoes, August 2016, http://www.ustornadoes.com/2016/08/26/surprise-indiana-ohio-tornado-outbreak-august-24-2016-happened/.

6   *The Arikara called it the Black Wind:* Nani Suzette Pybus, "Whirlwind Woman: Native American Tornado Mythology and Global Parallels" (PhD diss., Oklahoma State University, December 2009), 39, 43, 234.

7   *Researchers still dream of the day:* Interviews with Joshua Wurman, founder of the Center for Severe Weather Research.

7   *In Romania, where tornadoes are infrequent:* Bogdan Antonescu and Aurora Bell, "Tornadoes in Romania," *Monthly Weather Review*, March 2015, http://journals.ametsoc.org/doi/full/10.1175/MWR-D-14-00181.1.

CHAPTER ONE: THE WATCHER

11   *Fog clings to the low swells:* National Weather Service Daily Summary for Local Weather, July 21, 1993, http://maps.wunderground.com/history/airport/KAKO/1993/7/21/DailyHistory.html?req_city=Anton&req_state=CO&req_statename=&reqdb.zip=80801&reqdb.magic=1&reqdb.wmo=99999.

12   *The rain is coming down hard now:* Tim Samaras, *Driven by Passion*, DVD.

12   *It's isolated, rising above the cloud deck:* William Reid, "July 21, 1993, Last Chance, Colorado, Tornado," http://stormbruiser.com/chase/1993/07/21/july-21-1993-last-chance-colorado-tornado/.

15   *He has never chased outside his home state:* Andy Van De Voorde, "Swept Away: Storm Chasers Don't Need a Weatherman to Know Which Way the Wind Blows," *Denver Westword*, August 26, 1992, 26.

Chapter Two: A Boy with an Engineer's Mind

18   *The bane of his mother's household appliances:* Interviews with Jim Samaras, brother of Tim Samaras.

19   *Tim built his first transmitter:* Tim Samaras, WJ0G, QRZ.com, https://www.qrz.com/db/WJ0G.

19   *The house echoed with the roar:* Interviews with Jim Samaras.

20   *his first real glimpse:* Stefan Bechtel. *Tornado Hunter: Getting Inside the Most Violent Storms on Earth.* (Washington, D.C.: National Geographic Society, 2009), 47–49.

20   *For spending cash:* Interviews with Jim Samaras.

21   *this imperturbable faith:* Interviews with Larry Brown.

22   *Among his earliest projects:* "Minor Scale Event—Test Execution Report," Defense Nuclear Agency, January 30, 1986, 1.

22   *To track the blast's cratering characteristics:* Interviews with Robert Lynch.

22   *The day of the test:* Interviews with Larry Brown.

23   *Then, on a winter day in 1980:* Interviews with Kathy Samaras.

Chapter Three: This Love Affair with the Sky

27   *The urge returns the way it first began:* Bechtel, "Tornado Hunter," 9.

28   *In 1990 he enrolls in a six-week:* Interviews with Judi Richendifer.

28   *He learns why Tornado Alley is such a powder keg:* Conversations with Gabe Garfield.

29   *In the years before smartphones:* Interviews with Tim Tonge.

29   *antennas swaying like reeds from the roof of the Datsun:* Interviews with Brad Carter.

29   *Tim colonizes a used blue '91 Dodge Caravan:* Tim Samaras, "Dryline Chaser," StormEyes.org, http://www.stormeyes.org/tornado/vehicles/timsam.htm.

30   *Over at DRI, they call it kludging:* Interviews with Bud Reed.

30   *On storm days:* Interviews with Pat Porter.

30   *He remains a good husband:* Interviews with Kathy Samaras.

31   *"Some call it a hobby":* Van De Voorde, "Swept Away," 26.

31   *the Dryline Chaser is now recognizable:* Samaras, *Driven by Passion.*

32   *"I got some great stuff out in Kansas":* Interviews with Mike Nelson.

33   *When the spring and early summer have passed:* Interviews with Kathy Samaras.

33 *To Amy's profound embarrassment:* Interviews with Amy (Samaras) Gregg and Jenny (Samaras) Scott.

CHAPTER FOUR: THE SPARK

36 *Nobody saw it coming:* Interviews with Frank Tatom.

37 *The damage path was a half mile at its widest:* "November 15, 1989, Tornado Details," National Weather Service, https://www.weather.gov/hun/hunsur_1989-11-15_tornadodetails.

37 *the energy equivalent of half a ton of TNT per second:* Frank B. Tatom and Stanley Vitton, "The Transfer of Energy from a Tornado into the Ground," *Seismological Research Letters,* January 2001, fig. 2.

38 *Its components are quite simple:* Frank B. Tatom, and Stanley Vitton, "Method and Apparatus for Seismic Tornado Detection," US Patent and Trademark Office, January 1995.

39 *Tim is more excited than he has ever been:* Interviews with Pat Porter.

39 *Just south of Rome, Kansas:* Samaras, *Driven by Passion.*

41 *He's good at this:* Interviews with Frank Tatom and Pat Porter.

41 *What is stopping Tim:* Interviews with Pat Porter.

42 *The answer arrives in 1998:* Interviews with Larry Brown.

CHAPTER FIVE: CATCHING THE TORNADO

44 *a 1925 monster that left a three-state trail of destruction:* Peter S. Felknor, "The Tri-State Tornado," iUniverse, 2004.

44 *official policy forbade even the utterance:* Timothy A. Coleman and Kevin R. Knupp, "The History (and Future) of Tornado Warning Dissemination in the United States," *Bulletin of the American Meteorological Society,* May 2011, 569.

44 *stampeding cattle:* Marlene Bradford, "Historical Roots of Modern Tornado Forecasts and Warnings," *Weather and Forecasting,* August 1999, 508.

45 *It took two freak storms:* Robert A. Maddox and Charlie A. Crisp, "The Tinker AFB Tornadoes of March 1948," *Weather and Forecasting,* August 1999, http://journals.ametsoc.org/doi/full/10.1175/1520-0434(1999)014%3C0492%3ATATOM%3E2.0.CO%3B2.

45 *The first civilian tornado "bulletins":* Charles A. Doswell, Alan R. Moller, and Harold E. Brooks, "Storm Spotting and Public Awareness since the First Tornado Forecasts of 1948," *Weather and Forecasting,* August 1999,

http://journals.ametsoc.org/doi/full/10.1175/1520-0434(1999)014%
3C0544%3ASSAPAS%3E2.0.CO%3B2.

45   *The areas encompassed by any given watch were so vast:* Stephen F. Corfidi,
     "The Birth and Early Years of the Storm Prediction Center," *Weather and
     Forecasting,* August 1999, 515.

45   *Twisters seemed to be utterly repelled:* Howard B. Bluestein, *Tornado Alley:
     Monster Storms of the Great Plains* (Oxford: Oxford University Press,
     1999).

46   *Launched in 1945, the Thunderstorm Project:* Roscoe R. Braham, "The Thun-
     derstorm Project, 18th Conference on Severe Local Storms Luncheon
     Speech," *Bulletin of the American Meteorological Society,* August 1996, https://
     docs.lib.noaa.gov/noaa_documents/NOAA_related_docs/history/
     thunderstorms/thunderstorm.html.

48   *Each storm, Byers discovered, results from a confluence:* Interviews with Gabe
     Garfield.

49   *Keith Browning:* K. A. Browning and G. B. Foote, "Airflow and Hail
     Growth in Supercell Storms and Some Implications for Hail Sup-
     pression," *Quarterly Journal of the Royal Meteorological Society,* July
     1976, http://onlinelibrary.wiley.com/doi/10.1002/qj.49710243303
     /abstract.

49   *an updraft on steroids:* Keith A. Browning, "Airflow and Precipitation Tra-
     jectories within Severe Local Storms," *Journal of the Atmospheric Sciences,*
     November 1964, http://journals.ametsoc.org/doi/pdf/10.1175/1520-
     0469(1964)021%3C0634%3AAAPTWS%3E2.0.CO%3B2.

50   *In the turbulent springtime months:* Interviews with Gabe Garfield.

51   *The first Doppler scan of a tornado:* Rodger Brown, ed., "The Union City,
     Oklahoma, Tornado of 24 May 1973," NOAA Technical Memorandum
     ERL NSSL-80, December 1976, 3.

52   *they assumed they had found an error in the data:* Bluestein, *Tornado Alley,* 56.

53   *tornado got too close:* A. J. Bedard and C. Ramzy, "Surface Meteorolog-
     ical Observations in Severe Thunderstorms. Part I: Design Details of
     TOTO," *Journal of Climate and Applied Meteorology,* May 1983, 911.

54   *So while attending an after-hours party:* Interviews with Al Bedard and
     Howie Bluestein.

55   *Unless anchored or widened:* Howard B. Bluestein, "A History of Se-
     vere-Storm-Intercept Field Programs," *Weather and Forecasting,* August
     1999, http://journals.ametsoc.org/doi/full/10.1175/1520-0434%2819
     99%29014%3C0558%3AAHOSSI%3E2.0.CO%3B2.

56   *The instrument promptly pitched over onto its side:* TOTO, http://www.spc
     .noaa.gov/faq/tornado/toto.htm.

56 *"It would be fascinating to actually get inside":* A Day in the Life of a Storm
   Chaser, PBS, http://www.pbs.org/wgbh/imax/life.html.

57 *But this, too . . . failed to pierce the core:* Interview with William P. Winn,
   Langmuir Laboratory.

Chapter Six: The Cowboy Science

59 *So the answer he finally strikes upon:* Roy Heyman, "Air Blast Response of
   Low Drag Shape Launcher Vehicles," Small Business Innovation Research, Small Business Administration, 1991.

60 *quarter-inch-thick mild steel:* Timothy M. Samaras and Julian J. Lee, "Pressure Measurements within a Large Tornado" (conference paper, Eighth
   Symposium on Integrated Observing and Assimilation Systems for Atmosphere, Oceans and Land Surface, 2004).

60 *The project is an exceedingly unusual one:* Interviews with Larry Brown.

62 *The real selling point:* Rod Franklin, "Tornado Chasers: Engineer's Device Will Help Measure Their Fury," *Boston Business Journal,* http://www
   .bizjournals.com/boston/blog/mass-high-tech/2002/04/tornado
   -chasers-engineers-device-will-help.html.

63 *If they have learned anything from TOTO:* Interviews with Al Bedard.

64 *Tim's dream is made flesh:* "A Hardened In Situ Tornado Pressure Recorder," 1999, https://www.sbir.gov/sbirsearch/detail/93740.

64 *the turtle gets its first:* Interviews with Julian Lee.

65 *a fleet of the probes:* "A Hardened In Situ Tornado Pressure Recorder,"
   2000, https://www.sbir.gov/sbirsearch/detail/93742.

Chapter Seven: A Turtle in the Wild

66 *a credible plan to penetrate the tornado core:* Interviews with Anton Seimon.

68 *The mission's title underscores the danger:* National Geographic Society Expeditions Council Grant Application Form: "Inside Tornadoes: A Research Initiative."

69 *coordination issues dog the squadron:* Interviews with Anton Seimon.

69 *the classic plains setup:* Anton Seimon's personal chase log.

69 *he finds the arrangement chafing:* Interviews with Anton Seimon.

70 *Eleven days into the mission:* Ibid.

71 *It uprooted hardwoods:* Al Pietrycha, StormEyes.com chase summary, 2001,
   http://stormeyes.org/pietrycha/vortex/010618/chasesummary.txt.

Chapter Eight: The Toreador

74   *The pair now imagine a stripped-down mission:* Interviews with Anton Seimon.

74   *Tim believes he should be the one to make the call:* Email correspondence between Tim Samaras and Anton Seimon.

74   *a manifesto of sorts:* Mission outline written by Tim Samaras.

75   *Tim is chasing harder now:* Interviews with Pat Porter.

77   *Until Lee began chasing:* Interviews with Julian Lee.

77   *Tim misses an outbreak:* "Service Assessment: La Plata, Maryland, Tornado Outbreak," National Weather Service, September 2002, http://www .weather.gov/media/publications/assessments/laplata.pdf.

78   *closer than he's ever been:* Julian Lee and Tim Samaras, "Pressure Measurements within a Large Tornado" (American Meteorological Society conference paper).

78   *But the storm will not wait:* "Service Assessment: Veterans Day Weekend Tornado Outbreak of November 9–11, 2002," National Weather Service, March 2003, http://www.weather.gov/media/publications/assessments /veteran.pdf.

Chapter Nine: Stratford, Texas

80   *The road trip to the target:* Interviews with Anton Seimon.

82   *Clouds as dull as slag:* Chase video provided by Anton Seimon.

87   *In the paper that Tim coauthors with Wurman:* Joshua Wurman and Timothy Samaras, "Comparison of In Situ Pressure and DOW Doppler Winds in a Tornado and RHI Vertical Slices through 4 Tornadoes during 1996–2004" (conference paper, 22nd Conference on Severe Local Storms, American Meteorological Society, 2004).

Chapter Ten: Manchester, South Dakota

89   *The machine was to be one of a kind:* Interviews with William Gallus.

91   *Months later, in a cheap motel room:* Interviews with Pat Porter.

92   *This year, he was able to get the National Geographic Society:* Interviews with Rebecca Martin, National Geographic Society.

92   *Peter has already begged his editor for two extensions:* Interviews with Pat Porter.

93   *Rhoden and the others had pressed for a play:* Interviews with Gene Rhoden.

94   *its funnel is gracefully tapered:* Samaras, *Driven by Passion.*

95   *The twister has kept to the fields:* Ibid.

96   *The clouds are painted:* Ibid.

105 *That night, in a Huron, South Dakota, motel:* Footage provided by Gene Rhoden.

105 *With a back-of-the-envelope calculation:* Julian J. Lee, T. Samaras, and C. Young, "Pressure Measurements at the Ground in an F-4 Tornado" (conference paper, 22nd Conference on Severe Local Storms, 2004).

CHAPTER ELEVEN: DOUBLING DOWN

109 *Next he's on the CNN set:* YouTube clip of CNN segment with Soledad O'Brien, https://www.youtube.com/watch?v=4mKxPY6BiVc.

110 *Tim's probe has proven so capable:* Julian Lee and Tim Samaras, "Pressure Measurements at the Ground in an F-4 Tornado," https://ams.confex .com/ams/11aram22sls/techprogram/paper_81700.htm.

110 *Tim basks in recognition and praise:* Interviews with Julian Lee.

111 *The first hint of an answer:* Interviews with Bill Gallus.

114 *Tim is not just the leader:* Interview with Josh Wurman.

116 *Tim was irked:* Tim Samaras email correspondence.

117 *This sets off alarm bells:* Ibid.

CHAPTER TWELVE: A TEAM OF UPSTARTS

119 *Tim has in mind a woman:* Interviews with Cathy Finley and Bruce Lee.

123 *The other permanent member:* Interviews with Bob Young.

125 *a connection of an altogether different sort:* Interviews with Matt Winter.

127 *he seems a little chagrined:* Interviews with William Gallus.

CHAPTER THIRTEEN: TWISTEX TAKES THE GRAVEL ROAD

130 *He pulls Tim into a quiet room:* Interviews with Josh Wurman.

133 *Basically, "TWIST":* Tim Samaras email correspondence.

134 *Lee and Finley have standing invitations:* Interviews with Bruce Lee and Cathy Finley.

Chapter Fourteen: Quinter, Kansas

136 *From the lead car:* Chris Collura, May 23, 2008, chase video, https://www.youtube.com/watch?v=dikGjoPmeqI.

137 *The mesonets tack north along a sodden dirt road:* Ibid.

138 *That probably isn't good:* Interviews with Bruce Lee and Cathy Finley.

139 *"That whole thing's a tornado!":* Ibid.

139 *"We got a great view of this amazing":* Stefan Bechtel, "Tornado Hunter: Getting inside the Most Violent Storms on Earth," National Geographic Society, 2009, 91–93.

140 *Minutes later, on the outskirts of town:* Tim Samaras, *Driven by Passion II,* DVD.

143 *Karstens felt privileged:* Interviews with Chris Karstens.

144 *The night after Quinter, this potential is on everyone's mind:* Interviews with Bruce Lee and Cathy Finley.

145 *Storm chasers have practically commandeered Applebee's:* Interviews with Bruce Lee.

145 *He cues up his video of the day:* Doug Kiesling, May 23, 2008, chase video, https://www.youtube.com/watch?v=Fzk_vnGftYE.

146 *"somebody's gonna get bit":* Bechtel, "Tornado Hunter," 99.

Chapter Fifteen: "You Have My Only Son"

147 *In his eyes, Tim is a giant:* Interviews with Kathy Samaras.

147 *Beyond them, suspended mist:* Samaras, *Driven by Passion II.*

149 *he'd go see them alone the first time:* Interview with Jenny "Samaras" Scott.

150 *He takes photographs of tangerine sunsets:* Paul Samaras, Facebook page.

151 *The young man is thrilled:* Interviews with Kathy Samaras.

151 *Hail piles in pristine drifts:* Ed Grubb, May 15, 2010, chase video, Campo, Colorado.

Chapter Sixteen: Warnings

153 *The media probe logged a direct hit:* Bruce D. Lee, Catherine A. Finley, and Timothy M. Samaras. "Surface Analysis near and within the Tipton, Kansas, Tornado on 29 May 2008," *Monthly Weather Review,* February 2011.

154   *on the cusp of the 2009 season:* Interview with Rebecca Martin, National Geographic Society.

154   *For the moment:* Interviews with William Gallus.

154   *The instinctual calculus:* Ibid.

156   *In December 2008, the Discovery Channel reaches out:* Interviews with Bruce Lee and Cathy Finley.

156   *His . . . costars, Reed Timmer and Sean Casey:* Discovery Channel's *Storm Chasers.*

156   *The fifty-five-year-old pensioner:* Interviews with Ed Grubb.

157   *Under the watchful eye:* Interviews with William Gallus.

157   *The team surrounds the storm:* Karen Kosiba and Joshua Wurman, "Genesis of the Goshen County, Wyoming, Tornado on 5 June 2009 during VORTEX2," *Monthly Weather Review,* April 2013,

158   *Rare are the moments:* Interviews with Jenny Samaras Scott.

159   *new probe vehicle:* Jeff Glucker, "*Storm Chasers* Adds Super GMC Sierra to Tornado Chase Fleet," AutoBlog.com, November 14, 2010, http://www .autoblog.com/2010/11/14/storm-chasers-adds-super-gmc-sierra-to -tornado-chase-flee/#slide-265823.

159   *"What I'm hoping for":* Tim Samaras, *Storm Chasers,* season 4, episode 4.

Chapter Seventeen: Bowdle, South Dakota

161   *TWISTEX misses the year's first major tornado:* Interviews with Ed Grubb.

161   *They stop off in Minneapolis:* Interviews with Bruce Lee and Cathy Finley.

163   *Best of all, they learn that VORTEX2:* Interviews with Cathy Finley.

163   *Around four that afternoon:* Mesonet dash-camera footage provided by Bruce Lee and Cathy Finley.

165   *North of the mesonets:* Storm Chasers, season 4, episode 5.

166   *The circulation is closing in:* Samaras, *Driven by Passion II.*

167   *Off to the west:* Mesonet dash-camera footage provided by Bruce Lee and Cathy Finley.

168   *As they enter the outskirts:* Ibid.

170   *Even more striking:* Catherine Finley and Bruce Lee, "Mobile Mesonet Observations of the Rear-Flank Downdraft Evolution Associated with a Violent Tornado near Bowdle, SD, on 22 May 2010" (American Meteorological Society conference paper, 2010).

171   *Grzych and Carl's appraisal:* Mesonet dash-camera footage provided by Bruce Lee and Cathy Finley.

171   *The following day:* Storm Chasers, season 4, episode 5.

Chapter Eighteen: A Dead End, a New Chance

173   *He had always been a hit:* Interviews with Kathy Samaras.

174   *now too often congested with mesonets:* Ibid.

175   *Curiously, the camera operators:* Interviews with Bruce Lee and Cathy Finley.

175   *Aside from a few harmless dramatics:* Interviews with William Gallus.

176   *The premiere is a major event:* Interviews with Bob Young.

176   *funding seems to arrive at the last minute:* Interviews with Bruce Lee and Cathy Finley.

176   *after-action interviews:* Interviews with Bruce Lee, Cathy Finley, Ed Grubb, Tony Laubach.

177   *One evening, they attend a concert:* Interviews with Melissa June Daniels.

177   *When Lee and Finley approach:* Interviews with Bruce Lee, Cathy Finley, and Ed Grubb.

177   *The event that epitomizes:* Interviews with Ed Grubb.

178   *On April 27, near Aliceville, Alabama:* Ibid.

178   *Suddenly, they spot a tornado:* Tony Laubach, chase footage, April 27, 2011, https://www.youtube.com/watch?v=Ttni6owIWsw.

179   *No one argues when Grzych notes: Storm Chasers,* season 5, episode 1.

179   *Later that day, near Pleasant Ridge:* Interviews with Ed Grubb.

179   *By the September 25 premiere of . . . third season:* Interviews with Kathy Samaras.

180   *In early 2012, Tim phones Lee and Finley:* Interviews with Bruce Lee and Cathy Finley.

180   *Tim insists that TWISTEX will return:* Interviews with William Gallus.

181   *Out of work and now semiretired:* Interviews with Geoff Carter.

182   *Hurricane Isaac is already darkening:* Tim Samaras Twitter account.

183   *During some thunderstorms, electrical discharges:* Interviews with Walt Lyons.

184   *"When you approach eight hundred thousand":* Video of Tim Samaras in the LIV provided by Walt Lyons.

184   *"I want to see that formation process":* David Drummond, Dryline Media interview with Tim Samaras, February 2010, https://www.youtube.com/watch?v=FycVcmpYTfM.

184   *Soon, he and Paul are on the trail of a powerful EF4:* Tim Samaras chase video.

185   *he . . . gets about half of what he'd asked for:* Interviews with Bruce Lee and Cathy Finley.

Chapter Nineteen: Chase Nirvana

189 *For the first time since 2011:* Interviews with Ed Grubb.

189 *With the sun setting in the background:* Chase video, May 18, 2013, provided by Ed Grubb.

190 *It's like the early years:* Interviews with Ed Grubb.

190 *That night, the four check into a Comfort Suites:* Interviews with Ed Grubb, Marc Austin, and Sharon Austin.

191 *"You spend three or four days in a vehicle":* Jane J. Lee, "A Tornado Chaser Talks about His Science and Craft," News, National Geographic.com, May 22, 2013, http://news.nationalgeographic.com /news/2013/05/130520-tornado-chaser-samaras-thunderstorm -science/.

191 *Carl steers the probe truck:* Chase video, May 20, 2013, provided by Ed Grubb.

192 *"Careful! It's right next to us":* Ibid.

192 *"Wow! That was pretty exciting":* Paul Samaras chase video, provided by Kathy Samaras.

192 *Any grievance must seem trivial:* Interviews with Ed Grubb.

192 *Two elementary schools:* Erica D. Kuligowski et al., "Preliminary Reconnaissance of the May 20, 2013, Newcastle-Moore Tornado in Oklahoma," National Institute for Science and Technology Special Publication SP 1164, December 2013.

193 *what happened in Moore was no foregone conclusion:* Jon Erdman, "Moore, Oklahoma, Tornado 3 Years Later: What Turned It Violent?," Weather Channel, May 19, 2016, https://weather.com/storms/tornado/news/ moore-oklahoma-tornado-2013-research.

193 *That night, the crew busts on a storm:* Interviews with Ed Grubb.

194 *By the time they pass Robinson Avenue:* Paul Samaras chase video, provided by Kathy Samaras.

194 *If they could have gotten closer:* Kuligowski et al., "Preliminary Reconnaissance."

195 *The house is a flurry of activity:* Interviews with Kathy Samaras.

196 *On May 29, they settle around Salina:* Interviews with Walt Lyons.

197 *a familiar crew pulls up behind them:* Interviews with Bruce Lee and Cathy Finley.

197 *"We killed it":* Interviews with Cathy Finley and Bruce Lee.

198 *the three stop off in Alva:* Interviews with Walt Lyons.

Chapter Twenty: A Shift in the Wind

199   *At around four that afternoon:* Interviews with Rick Smith.

200   *a nerve-racking twelve-hour shift:* Interviews with Marc Austin.

Chapter Twenty-One: El Reno, Oklahoma

205   *A white Chevy Cobalt:* Interviews with Gabe Garfield.

206   *The dire language:* "May 31, 2013, 2000 UTC Day 1 Convective Outlook," Storm Prediction Center, National Weather Service, http://www.spc .noaa.gov/products/outlook/archive/2013/day1otlk_20130531_2000 .html.

206   *The values this afternoon are astonishing:* Interviews with Jeff Snyder.

207   *At 5:41 p.m.:* Timeline provided by Gabe Garfield.

207   *The distended wall cloud pulses:* El Reno Tornado Environment Display, courtesy of Anton Seimon, http://el-reno-survey.net/ted/.

207   *Over . . . twenty minutes they cover roughly three miles:* Interviews with Gabe Garfield.

207   *Tim's face appears in profile:* Video stills, portions of video transcript, and timeline from 2014 National Storm Chasers Convention presentation by Gabe Garfield.

208   *"Oh, my God":* Ibid.

208   *Within seconds of the first wisps, there is a tornado on the ground:* El Reno Tornado Environment Display.

208   *"My God . . . Wow, look at the tornado! Just to our south":* 2014 National Storm Chasers Convention presentation by Gabe Garfield.

208   *Instead, it seems to be receding:* El Reno tornado-track timeline, width, radar centroids, via Google Earth, courtesy of Gabe Garfield.

208   *They continue south for a mile:* Interviews with Gabe Garfield.

209   *"Is the airport down another mile?":* 2014 National Storm Chasers Convention presentation by Gabe Garfield.

210   *At 6:04 p.m.:* Interviews with Jeff Snyder and Howie Bluestein.

210   *Nearby, his RaXPol mobile radar generator thrums:* El Reno tornado footage from RaXPoL, courtesy of Jeff Snyder, https://www.youtube.com/ watch?time_continue=1&v=EFwJAlg0qWc.

211   *Bluestein . . . only two options:* Interviews with Howie Bluestein and Jeff Snyder.

CHAPTER TWENTY-TWO: THE DRAGON'S TAIL

213 *At 6:12 p.m. they've finally arrived at their east route:* Interviews with Gabe Garfield.

214 *"Okay, we've gotta be careful in case this thing wraps up":* Interviews with Gabe Garfield; 2014 National Storm Chasers Convention presentation by Gabe Garfield; and Robert Draper, "Last Days of a Storm Chaser," *National Geographic*, November 2013, http://ngm.nationalgeographic.com/2013/11/biggest-storm/draper-text.

215 *"We've got debris in the air":* 2014 National Storm Chasers Convention presentation by Gabe Garfield.

215 *Tim sets the camera down:* Interviews with Gabe Garfield.

215 *Directly to their south:* El Reno Tornado Environment Display.

215 *At 6:16 p.m., Josh Wurman:* Joshua Wurman et al., "The Role of Multiple-Vortex Tornado Structure in Causing Storm Researcher Fatalities," *Bulletin of the American Meteorological Society,* January 2014.

216 *Wurman and his colleague Karen Kosiba:* Interviews with Josh Wurman.

216 *Much of what follows will be discovered:* Wurman et al., "Role of Multiple-Vortex."

217 *Those questions will come later:* Interviews with Josh Wurman.

CHAPTER TWENTY-THREE: THE CROSSING

218 *The white Cobalt . . . shoulders through an inflow current:* Interviews with Gabe Garfield.

218 *"Now we go up north . . . then east":* Ibid.

219 *"the tornado is about five hundred yards away":* 2014 National Storm Chasers Convention presentation by Gabe Garfield; and Draper, "Last Days of a Storm Chaser."

221 *The last update refreshed nearly five minutes earlier:* Interviews with Gabe Garfield.

221 *US 81 disappears:* El Reno Tornado Environment Display.

221 *"So, this is the highway":* 2014 National Storm Chasers Convention presentation by Gabe Garfield.

222 *Due south of the intersection:* Mike Bettes, Weather Channel live report, May 31, 2013, https://www.youtube.com/watch?v=m4WfuhUrWHs.

223 *Bettes rides in the front passenger seat:* Interviews with Austin Anderson.

223   *The blinding rain is the first sign:* Chase video from inside the SUV provided by Austin Anderson.

224   *The bus driver:* Nolan Clay, "El Reno Tornado Survivor Tells of Others' Deaths," *Oklahoman,* June 7, 2013, http://newsok.com/article/3842601.

Chapter Twenty-Four: The Last Ride

225   *Carl races across:* Rear-dash-camera footage provided by Dan Robinson.

225   *the radar is misleading:* Interviews with Gabe Garfield.

226   *"Now I see it":* Ibid.

226   *"In fact, uh, keep going":* 2014 National Storm Chasers Convention presentation by Gabe Garfield.

226   *headwinds upward of 110 miles per hour:* Interviews with Gabe Garfield.

227   *But we do know what happens inside the storm:* Wurman et al., "Role of Multiple-Vortex."

Chapter Twenty-Five: How Far from Daylight

229   *Dan Robinson knows nothing of the headlights:* Interviews with Dan Robinson.

231   *Moments after the subvortex enters Reuter Road:* Interviews with Howie Bluestein and Jeff Snyder.

231   *The tornado has transformed dramatically:* El Reno tornado footage from RaXPoL, courtesy of Jeff Snyder, https://www.youtube.com/watch?v=FrfAfJNBT3A.

Chapter Twenty-Six: Ground Truth

233   *Shortly before 7:00 that evening:* Interviews with Doug Gerten.

234   *"signal 30":* Canadian County Sheriff's Office Radio Call Log, May 31– June 1, 2013.

235   *He snaps a picture . . . will soon spread:* Interviews with Gabe Garfield.

235   *Too restless to sleep:* Interviews with Marc Austin and Sharon Austin.

236   *The day after:* Interviews with Kathy Samaras.

236   *Tuesday morning, Garfield and Fox return to Reuter Road:* Interviews with Gabe Garfield and Erik Fox.

239   *the details are grounded in fact but open to interpretation:* Interviews with Kathy Samaras.

CHAPTER TWENTY-SEVEN: THE SIGNS

CHAPTER TWENTY-EIGHT: TIM'S LEGACY

EPILOGUE

INDEX

ABOUT THE AUTHOR

BRANTLEY HARGROVE is a journalist who has written for *Wired*, *Popular Mechanics*, and *Texas Monthly*. In his reporting, he has explored the world of South American jewel thieves who terrorize diamond dealers in South Florida. He's gone inside the effort to reverse-engineer supertornadoes using supercomputers. And he has chased violent storms from the Great Plains down to the Texas coast, including a land-falling Category 4 hurricane and one of the rarest tornadic events in recent memory: twin EF4 tornadoes that chewed through a small Nebraskan farming village. He lives in Dallas, Texas, with his wife, Renee, and their two cats. *The Man Who Caught the Storm* is his first book.

4-18

MF